TEETH SADDLE COINS AQUEDUCT ARCH STIRRUPS ASPIRIN CATAPULT PA

MACHINE DOME FORK CAROUSEL HORSESHOES WINDMILL PAPER MONEY

PRINTING PRESS GLOBE SCREW SAWMILL NEWSPAPER TELESCOPE BAROM

TOILET FRANKLIN STOVE GLUE LIGHTNING ROD BIFOCALS JIGSAW PUZZL

TON BALL BEARINGS PARACHUTE LOCOMOTIVE MATCHES TIN CAN STETHO

ICE COLT REVOLVER MORSE CODE BICYCLE CAMERA POSTAGE STAMP ANES

SEPTICS ODOMETER SAFETY PIN MILKING MACHINE ELEVATOR SYRINGE L

GROUND TRAINS DYNAMITE PERIODIC TABLE OF THE ELEMENTS SQUARE-BO

NAL COMBUSTION ENGINE TELEPHONE MICROSCOPE PHONOGRAPH REFRIGE

SCRAPER AUTOMOBILE DISHWASHER CONTACT LENSES MOTION PICTURES ES

O X-RAY MOUSETRAP HEARING AID PAPER CLIP WASHING MACHINE AIR C

RED CRAYONS WINDSHIELD WIPER FLY SWATTER OUTBOARD MOTOR PLAST

ALS ARMORED COMBAT TANK WRISTWATCH DRYWALL ADHESIVE BANDAGE

MACHINE LIQUID-FUELED ROCKET AEROSOL SPRAY RESPIRATOR TELEVISION

MACHINE NYLON PARKING METER RADAR PHOTOCOPIER SUNSCREEN ALU

T PEN HELICOPTER PAPERBACK BOOK DUCT TAPE AQUALUNG ATOMIC BOMB

MAKER CREDIT CARD DISPOSABLE DIAPERS BAR CODE FIBER OPTICS CALCU

L SATELLITE LASER THREE-POINT SEAT BELT ULTRASOUND IMAGING IN VITI

HINE) AIR BAGS COMPUTER MOUSE SMOKE DETECTOR COMPACT DISK EMAI

BAL POSITIONING SYSTEM (GPS) THE INTERNET DNA FINGERPRINTING PAINT

K LOOM SAUNA WHEEL WIGS SUNDIAL PLOW BRONZE OVEN ALPHABET

S GLASS RIVET SOAP PORCELAIN SCISSORS CARTS AND CHARIOTS FALSE TE

THOUSE WALLPAPER CLOTHES IRON CEMENT COMPASS EASEL VENDING M

POWDER FIREWORKS AMBULANCE CANNON FLYING BUTTRESS HOURGLASS

TION LOCK MICROSCOPE STEAM ENGINE THERMOMETER DENTAL BRACES

NE BALLOON GUILLOTINE BATTERY COTTON GIN CORKSCREW PENCIL VA

EXTINGUISHER ELASTIC FABRIC BRAILLE LAWNMOWER MECHANICAL REAP

THE BOOK OF INVENTION

Thomas J. Craughwell

Tess Press

600
Cra

Special thanks to Stratford/Tex Tech Publishing Services and Charles Merullo, Jennifer Jeffrey, and Kate Pink at Endeavour.

The Book of Invention was created by Black Dog and Leventhal in conjunction with Endeavour London Limited.

<u>PHOTO CREDITS</u>

Granger: front cover, 1, 157, 291, 303; **HouseofAntiqueHardware.com:** front cover, 3; **LOC:** 7, 8, 9, 10, 14, 44, 77, 127, 146, 152, 153, 160, 161, 163, 171, 178, 179, 186, 191, 192, 196, 203, 211, 214, 218, 223, 230, 235, 236, 241, 244, 246, 252, 256, 262, 273, 281, 285, 287, 288, 299, 300, 304, 305, 310, 311, 315, 317, 322, 324, 328, 333, 336, 342, 351, 353, 366, 371, 376, 379, 382, 386, 396, 411, 419, 421, 436, 438, 455, 483; **Shutterstock:** front cover, 7, 8, 10, 11, 16, 18, 20, 22, 24, 27, 28, 30, 34, 38, 42, 46, 48, 54, 56, 58, 60, 64, 68, 73, 80, 83, 84, 90, 98, 100, 102, 106, 109, 110, 112, 116, 118, 120, 122, 125, 130, 132, 136, 142, 148, 154, 156, 158, 162, 165, 168, 170, 177, 189, 190, 194, 198, 201, 204, 206, 208, 210, 212, 220, 233, 240, 248, 250, 255, 260, 264, 267, 270, 272, 275, 278, 284, 292, 293, 296, 297, 303, 307, 308, 312, 316, 327, 331, 334, 337, 338, 341, 350, 352, 354, 356, 361, 362, 365, 368, 372, 373, 374, 378, 381, 384, 388, 400, 408, 407, 412, 416, 414, 422, 424, 428, 430, 434, 440, 444, 446, 450, 452, 454, 456, 458, 460, 462, 468, 469, 470, 473, 474, 478, 482, 487, 492, 494, 495, 499, 506, 512.

All images below courtesy of **Getty Images** including the following which have additional attributions:

Agence France Presse: 5(2L), 6(2L), 35, 47, 61, 71, 97, 472, 497, 507; **Altrendo:** 443; **Bridgeman Art Library:** 5(R)/Bibliotheque Nationale, Paris; 6(3L) Private Collection; 7(2L)/Private Collection; 7(3L)/Musee de la Revolution Francaise, Vizille, France; 21/Private Collection; 25/Great Palace Mosaic Museum, Istanbul; 31/Museo Nacional Palacio de B.Artes, Havana; 39/National Museum, Damascus; 45/British Museum, London; 49/Musee des Antiquites Nationales, St Germain-en-Laye, France; 55/Hamburger Kunsthalle, Hamburg; 59/Bibliotheque Nationale, Paris; 67/Harewood House, Yorkshire, UK; 69/Galerie Janette Ostier, Paris; 79/Private Collection; 81/Musee d'Histoire de la Medecine, Paris; 87/National Gallery, London; 89/Bibliotheque des Arts Decoratifs, Paris; 93/Musee des Beaux-Arts, Valenciennes, France; 95/British Museum, London; 99/Biblioteca Ambrosiana, Milan; 103/Private Collection; 113/Musee des Beaux-Arts, Rouen; 119/Ashmolean Museum, University of Oxford, UK; 121/Private Collection; 137/Bibliotheque Nationale, Paris; 141/Museum of London; 145/Bibliotheque Nationale, Paris; 155/Scott Polar Research Institute, Cambridge, UK; 159/Private Collection; 167/Private Collection; 183/Musee de la Revolution Francaise, Vizille, France; 193/Musee de la Chartreuse, Douai, France; 295/Royal Albert Memorial Museum, Exeter, Devon, UK; **de Agostini Picture Library:** 91; **Imagno:** 280; **Isifa:** 17; **Liaison:** front cover(2R), 1(R), 505, 511; **National Geographic Society:** 6(R), 23, 43, 53, 101, 111, 117, 475; **Popperfoto:** 397, 399, 449; **Roger Viollet:** 139, 147, 257, 279, 319; **Time and Life Pictures:** front cover (2L), back cover, 1(L), 5(L), 5(3L), 9(L), 9(R), 10(3L), 11(L), 15, 51, 85, 169, 173, 175, 185, 187, 195, 229, 234, 239, 249, 259, 283, 290, 309, 343, 357, 359, 363, 391, 395, 405, 410, 417, 427, 429, 431, 439, 441, 447, 453, 459, 461, 465, 485, 493, 503, 513.

This edition published by Tess Press, an imprint of
Black Dog & Leventhal Publishers, Inc.
151 West 19th Street
New York, NY 10011

Cover and Interior Design by Lindsay Wolff

Manufactured in China

ISBN-13: 978-1-60376-039-3

h g f e d c b

CONTENTS

ca. 30,000 B.C.

Paint

5000 B.C.

Loom

3000 B.C.

Alphabet

3000 B.C.

Buttons

3000 B.C.

Ink

3000 B.C.
Paper

297 B.C.
Lighthouse

10 B.C.
Compass

900
Fireworks

1456
Printing Press

1738
Flush Toilet

1789
Guillotine

1797
Parachute

1824
Braille

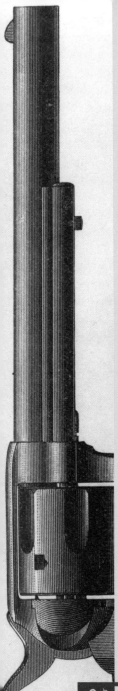

1836
Colt Revolver

1839
Bicycle

1845
Rubber Band

1854
Lifejacket

1874
Barbed Wire

1878
Light Bulb

1901
Air Conditioning

1903
Colored Crayons

1907
Outboard Motor

1913
Stainless Steel

1914
Turn & Brake Signals

1917
Adhesive Bandage

1927
Aerosol Spray

1939
Paperback Book

1942
Duct Tape

1945
Microwave Oven

1954
Robot

1955
Velcro

1972
Space Shuttle

1984
DNA Fingerprinting

INTRODUCTION

U.S. Patent office, 1869

As of this writing, the United States Patent Office has issued patents for more than four million inventions. That figure is an astonishing testimony to American ingenuity. Yet in all likelihood, very few of those patented inventions will make a major impact on our day-to-day life in the way that the printing press, the steam engine, the telephone, the automobile, and the Internet have done. And that question—Which inventions have been essential, or most useful, or have become ubiquitous?—served as the basis for selecting the 250 inventions featured in this book.

The first inventive minds were our prehistoric ancestors who discovered how to use the wheel, make pottery, and mix natural dyes and pigments for paint. All of these were advances in technology—although ancient people might not have expressed it that way. Nonetheless, through luck or accident, observation of forces in nature or the inspiration of the moment, or perhaps trial and error, these early inventors took what we would describe as scientific principles and found practical applications for them. Inventors have been doing the same ever since.

One of the most fascinating aspects of studying the history of invention is discovering how one invention led to others. For example, the first people to use the wheel must have noticed that it rolled much more smoothly over level ground—an observation that led to the construction of the first roads. The Romans are famous for their network of 50,000 miles of paved roads that stretched from the Eternal City to every corner of the empire. Perhaps even more impressive is the highway system of ancient China, where the roads were eight lanes across, with a ninth lane running down the center reserved for the exclusive use of the emperor and his family (think of it as the first express lane). In the same way, the history of communication begins with the invention of the alphabet, moves on through the invention of the telegraph and the telephone, and comes into our own time with the invention of e-mail in 1971.

Some inventions are so momentous they change everything forever. Thomas Newcomen's 1705 steam engine gave us a simple source of power that, for the first time in history, did not require human or animal muscle, nor was it dependent on the strength of a river's current or the stiffness of the breeze. The transistor invented in 1947 by John Bardeen and Walter Brattain (with improvements introduced by William Shockley) has been essential to the circuitry of virtually every electronic device invented ever since. And Timothy Berners-Lee's creation of the World Wide Web in

1983 has given anyone with access to a computer terminal admittance into a universe of knowledge that no brick-and-mortar library could ever hope to replicate.

Behind every invention in this book is a story of men and women with a deep grasp of science; a genius for mechanics; a high tolerance for setbacks, disappointments, and failure; and the perseverance to finally realize their dreams. Tragically, some of these brilliant minds failed to take basic precautions such as getting a patent for their invention or renewing their patent. Mary Anderson, who invented the automobile windshield wiper in 1904, let her patent expire. When the auto industry made windshield wipers standard equipment, they were not obliged to pay Anderson any royalties. She could have lived and died a millionaire; instead, Anderson spent her life as the owner and manager of an apartment building in Birmingham, Alabama.

Mary Anderson's story brings us to another remarkable facet in our story— inventions that went on the market even though they were incomplete (from our standpoint at least). The first American automobiles, for example, rolled out the factory door in Lansing, Michigan, 1901, without brake lights, no direction indicators, not even windshield wipers. These "accessories" simply hadn't occurred to the inventor or the mechanics that assembled the cars. We've already learned about Mary Anderson's windshield wipers. It was Florence Lawrence, a star of silent films and an ardent fan of automobiles—she was one of the first women in Hollywood to drive a car—who invented directional and brake indicators in 1914. While Anderson patented her invention but let the patent run out, Lawrence never patented her inventions at all. The automobile industry snapped them up without paying her a penny.

Of course, even inventors who did patent their inventions fell victim to unscrupulous competitors. Eli Whitney's cotton gin, Elias Howe's sewing machine, and Margaret Knight's machine to make a square-bottomed paper bag are just a few examples of inventions that were pirated. To defend their patents, many inventors spent long years and thousands of dollars in court. Whitney got so sick of trying to protect his patent that he gave up and started a new career designing factories. The story has a happier ending in the case of King C. Gillette, inventor of the disposable safety razor. The millions he made from sales of his razors gave him more than enough cash to buy out competitors who were infringing on his patent.

The Book of Invention offers an introduction to the men and women whose ingenuity made daily life a little easier, work more efficient and productive, transportation faster and more comfortable, and health care more effective. Their "eureka" moments changed the world.

Paint

The prehistoric cave paintings of France, such as those found at Lascaux, preserve the oldest surviving examples of paint. Analysis of these umber and ochre pigments revealed that the ancient artists used iron oxide and manganese oxide. Tens of thousands of years later, as civilizations sprang up in China, India, Egypt, and the Near East, bright blues, reds, yellows, and greens became available, but at a price—they were all derived by grinding up semi-precious stones, specifically lapis lazuli, cinnabar, orpiment, and malachite. By the first century A.D., white lead was available.

The initial recipe for paint was simple: The three components were a pigment, a binder such as egg, and a thinner that made the resulting mix easy to apply to a surface.

Oil paint, which used oil as a binder, was perfected in the fifteenth century in northern Europe. The first great practitioners of oil painting, Jan van Eyck, Rogier van der Weyden, and Hans Memling—all from what is now Belgium—set off a renaissance comparable to what was happening at the same time in Italy. When pigment was mixed with oil, the colors took on a translucent character unlike anything available previously. Van Eyck and his colleagues learned to apply coats of oil paint in thin layers that gave the colors an intense, brilliant glow. Until the nineteenth century, paint was the term reserved for the beautiful colors created by mixing pigments with oil; those that used a different medium such egg were know as distemper.

Bark, berries, and bugs have also produced paints: ground up buckthorn berries produce a shade known as Dutch pink (which is actually a shade of yellow); the dried body of a female cochineal beetle from Chile produces a bright red known as cochineal; and a bright green could be made by using arsenic.

For most of paint's history there was no such thing as ready-to-use paint; all paint, whether for a portrait or for the exterior of a house, was mixed on the spot by hand by trained masters. That changed in 1878 when two Americans, Henry Sherwin and Edward Williams, began to manufacture the first ready-mixed paints in resealable tin cans.

30,000 B.C.

A portion of the prehistoric cave painting in Lascaux, France

POTTERY

For many years, archaeologists believed pottery had originated in the Near East about 5000 B.C. Then, in the 1960s, archaeologists discovered pottery that dated to 11,000 B.C. on the island of Honshu in Japan. Basic pottery is wet clay that has been shaped, left to dry, and then fired to make it hard. If the object is covered with a liquid glaze, the firing process will harden and bind the glaze to the surface of the pot, making it waterproof.

If we expand the definition of pottery beyond clay pots and bowls to include any clay object that has been shaped and fired, then the oldest piece of pottery comes from Dolní Věstonice in the Czech Republic and dates from somewhere between 29,000 and 25,000 B.C. This pottery sculpture of a woman with wide hips and large pendulous breasts is known as "Venus of Dolní Věstonice." It stands 4.4 inches high and 1.7 inches wide and was discovered during an archaeological dig in 1925. On the figure is a child's fingerprint—the kid must have picked up and played with the soft clay figurine before it was fired.

The earliest pieces of pottery were shaped roughly by hand then fired in huge bonfires. Pottery wheels were not invented until about 5000 B.C. in Mesopotamia; kilns for firing pottery were in use in Persia and Mesopotamia around 4000 B.C. The potter's wheel and the kiln came at just the right moment: Previously, individuals made clay pots and bowls for their own use or for others in their tightly knit band of hunter-gatherers. By 5000 B.C., people were living in villages and cities, which gave potters a large market for their wares. With a wheel and a kiln, the potter could turn out a higher-quality product at a greater rate of speed.

29,000 B.C.

The Venus of Dolni Vestonice

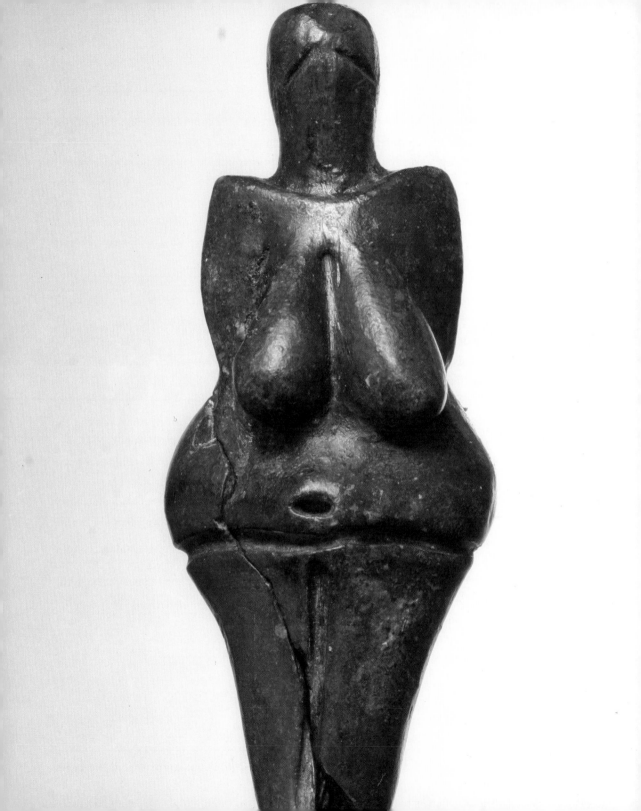

SEWING NEEDLE

Among the very first human tools were hunting weapons—spear heads, knife blades, and choppers for bringing down game and cutting up the carcass. However, at least 25,000 years ago our prehistoric ancestors were also making more delicate tools—clothes fasteners, awls to punch holes in leather, and sewing needles fashioned from animal bone. The oldest sewing needle discovered to date was found in southwest France and dates to about 23,000 B.C., while remnants of animal skins that had been tailored and sewn into pants, shirts, and shoes were found in a 22,000-year-old grave near Moscow in Russia.

Bone needles, however, are brittle. Around 5000 B.C., needles were made of bronze, a much stronger material. Needles made of bronze, brass, iron, and even of silver were the norm into the early Middle Ages. After the eighth century A.D., the Moors, who introduced steel to Spain, began making steel needles, as well as steel swords. Ladies kept their sewing needles in a special case, which many women wore dangling from their belts.

The typical needle used for all-purpose sewing is known as a *sharp*. As the name implies, it is sharp at one end; at the other end is the opening known as the eye through which the thread is passed. A short version, often called a quilting needle, is not only used for quilting, but also for fine sewing work such as tailoring clothes. A crewel needle, used for embroidery, has a larger eye than a sharp to accommodate the thicker yarns and threads used in embroidery. There are also long, heavy upholstery needles necessary for piercing heavy fabrics: Some upholstery needles are a foot long.

One necessary accessory for sewing—especially when sewing thick fabrics—is the thimble, typically a bell-shaped metal protector worn over a finger to push the needle through cloth or leather. Archaeologists discovered the oldest thimble so far in Pompeii, which dates no later than 79 A.D.

23,000 B.C.

The sewing needle was one of the first human tools

BOOMERANG

Ten thousand years ago, inhabitants of Egypt, Eastern Europe, and of course Australia used curved flat sticks we call boomerangs for hunting and fighting. The oldest boomerang found to date was discovered in a cave in the Carpathian Mountains in southern Poland; it is 28 inches across, and has been carbon dated to about 19,000 B.C. Most likely, the boomerang was used to hunt small prey such as rabbits and birds. It would have been used for a time in warfare, but after the invention of spears and bows and arrows, the boomerang retired from the battlefield and became used exclusively for hunting. There were also heavy boomerangs that could take down large animals, such as a kangaroo, but those boomerangs did not return to the thrower.

Boomerang is an Australian Aboriginal term that means "a curved blade." In other societies, it was known as a throwing club, a throwing stick, or a hunting stick. Because of its curved shape, it rotates in the air when thrown. A person throws the boomerang overhand, like a baseball. Before releasing the boomerang, the thrower snaps his wrist. The rotation of the boomerang in midair brings it back toward the thrower, however the flight path of the boomerang is not consistent so the thrower must watch to see where the boomerang will return. Since it is a weapon—and being struck by it will cause injury—the safest way to catch a returning boomerang is to "clap" it between the palms of both hands.

Unlike other weapons, caring for a modern boomerang is simple. It needs to be stored in a place where it can lay flat. If the boomerang has been nicked or scratched, fill the defect with wood putty, sand it down, and give the boomerang a fresh coat of polyurethane.

19,000
B.C.

Boomerang is Aboriginal for "curved blade"

Oil Lamps

Anyone who has barbecued has noticed that when meat fat drips onto the coals, the fire flares up. Our prehistoric ancestors noticed the same thing as they cooked their meat over an open flame. We now know that they collected fat, soaked a piece of moss or some other natural porous material in the drippings, placed it in a hollow bit of rock, and made the world's first oil lamp. Exactly when this occurred is open to debate—some historians believe oil lamps illuminated prehistoric cave dwellings 70,000 years ago. A more conservative estimate is that oil lamps were in use around 10,000 B.C.

By 1000 B.C., oil lamp technology had improved. Many societies used a dried rush as a wick, and they had learned that the flame burned more evenly if the wick rested in a groove slightly above the surface of the bowl of oil. By this time, meat drippings had been replaced by olive oil in the Mediterranean region; by ghee, or clarified butter, in India; and in Persia, people collected petroleum where it pooled on the ground. Lamps also became more decorative—even plain terra-cotta oil lamps, such as can be seen to this day in the Roman catacombs, were made in an appealing boat-shape, while lamps made of bronze or silver were often genuine works of art. By the fourth century A.D., lamps had moved out into the streets: The biblical scholar St. Jerome (345–420) reports that at night Antioch in Syria was lit by oil lamps suspended above the streets.

In the early years of the nineteenth century, whale oil became the primary lighting source for many homes and businesses. The "oil" is really a liquid wax derived from the blubber of the sperm whale. Whale oil, which sold for about $2 a gallon in the first half of the nineteenth century (about $200 in modern money), built countless fortunes in seaport towns along the Atlantic coast of the United States. After 1850, whale oil was supplanted by the cheaper kerosene, which some households still keep on hand in case the electricity goes out.

10,000

B.C.

Roman-era oil lamp shaped like a human head

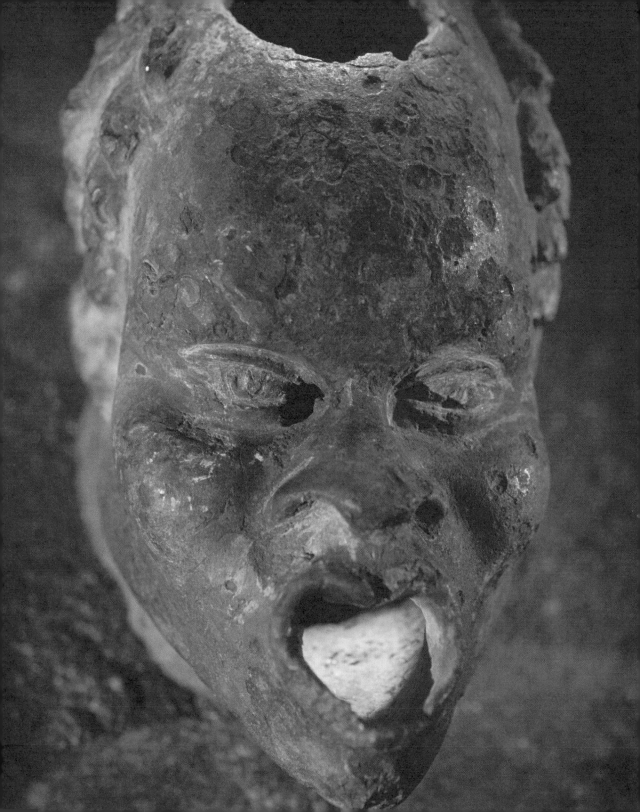

Baskets

It's likely that the first thing prehistoric people wove were mats to cover the floors of their huts and tents to protect sitters from the damp and chill of the bare earth. These mats were made of rushes, grasses, or strips of bamboo or wicker—in other words, whatever plants were available in the neighborhood. Once humans had learned how to weave mats they began to experiment with using the same materials to weave containers, and these containers were the first baskets.

Basket weaving appears to have sprung up in human societies all across the globe at about the same time; consequently, it is impossible to say where the first basket was born. Its arrival, however, was especially timely. Humans were living together in communities and they needed containers they could bring with them when collecting nuts, berries, and roots, harvesting grain, or bringing home the day's catch of fish. Containers were also essential for storing grain and seeds. Baskets were especially well suited to all of these needs—they were light and portable and, unlike clay pots, they did not break.

Although baskets are utilitarian objects, basket weaving was soon elevated to an art form. To this day, superb examples of complex weaving designs are still created by Native American, African, and Asian basket weavers. Beginning in the seventeenth century, settlers on the island of Nantucket began creating especially fine baskets by splitting ash, oak, and hickory. The Shakers, a religious group that thrived in the eastern United States during the nineteenth century, learned the basics of basket making from the Algonquin Indians with whom they traded. The Shakers improved upon the Indians' method by carving wooden molds in a variety of sizes and styles around which the baskets were woven. To their surprise, the Shakers found a huge market for the little "fancy baskets," as they called them. More decorative than useful, these pretty baskets were snapped up by non-Shaker consumers.

Today, Nantucket and Shaker baskets are displayed in folk art museums.

8000

B.C.

BRICKS

Workers shape and dry bricks in Jamestown, the first permanent English colony in America

Of all the various construction materials, bricks are arguably the best: the necessary materials—clay, water, and fire—are found virtually anywhere; bricks can be made into almost any size; a building built of brick is strong, durable, and will withstand all types of bad weather; and bricks provide an excellent source of insulation (which explains why most chimneys are constructed of brick).

Some 9,000 years ago, people in the Middle East used brick to build the walls of Jericho; 5,000 years ago, the Babylonians built their city of brick; and 2,300 years ago, the Chinese used brick to construct their Great Wall. The ancient Romans liked their bricks large and thin, set between thick layers of mortar. The English of the Renaissance period liked to lay their bricks in decorative patterns, the most common type being English Bond, in which a wall is constructed of alternating rows of short bricks called headers and bricks that are twice as long known as stretchers.

As for the brick-making process, it has barely changed in 9,000 years. First, the clay is blended with water and sand; pebbles, bits of plants, and other unnecessary materials are removed from the mixture; then the wet clay is pressed into a brick mold and set out to dry. The dry clay is fired in a kiln—the longer a brick is baked and the closer it is to the fire, the darker its color. Since the late nineteenth century, the mixing, cleaning, molding, drying, and firing processes have all been mechanized. In one respect, however, bricks have not changed at all: They are still light enough to be picked up easily and small enough so the bricklayer can wield his trowel with one hand and set the brick in place with the other.

7000 B.C.

The Great Wall of China is one of the most famous structures made from bricks

MIRROR

The first mirrors were the still surfaces of a pond or stream where ancient humans could gaze at their reflections. Greek mythology tells how an exceptionally good-looking young man named Narcissus became infatuated with his reflection in a pond, reached down to embrace it, tumbled into the water, and drowned.

In the 1960s, archaeologists discovered the world's oldest mirrors—highly polished pieces of obsidian, an extremely hard volcanic glass—in the graves of ten women at Çatal Hüyük in central Turkey. By 2900 B.C., the Egyptians used mirrors of polished bronze or copper. Similar mirrors were made in India and Pakistan around 2800 B.C. The most beautiful polished bronze mirrors were created in China during the Han Dynasty (202 B.C. to 220 A.D.).

The first compact was made in Greece during the fifth century B.C.: a small box that flipped open to reveal a polished metal mirror. Since metal tarnishes easily, some of these compacts also contained a sponge dipped in powdered pumice to keep the mirror clean and bright.

The Roman emperor Domitian, convinced that his enemies were plotting to murder him, covered the area where he exercised daily with mirrors so he could see assassins coming from any direction. Domitian's paranoia was justified—people were indeed out to get him. However, they left him alone in his private gym and killed him instead in his bedroom.

By the third century, the Romans had begun making crude glass mirrors by painting a thin layer of silver, gold, or copper on one side of a sheet of transparent glass. The image was wobbly, but it was a start. The craft was lost, however, during the barbarian invasions that destroyed the empire. Glass mirrors did not return until the thirteenth century.

Modern mirrors are made with a silvering process. A sheet of glass is placed in a vacuum chamber where heated aluminum atoms are blasted at the glass. As soon as they strike the glass the atoms cool and stick, creating a perfect reflective surface.

The main mirror for the Hubble Space Telescope

Sugar

Sugar cane, the source for about 70 percent of the world's sugar, originated in New Guinea 8,000 years ago. It did not travel far from its home turf until the year 1000 B.C., when merchants from India and Southeast Asia started bringing it home with them. Before the arrival of sugar, honey was the ubiquitous sweetener around the globe. It was chefs in India who began extracting sugar by boiling the cane juice. The grains of this sugar would have been brown in color.

Sugar did not reach the Mediterranean world until the seventh century A.D., during the Arab conquest of North Africa and the Iberian Peninsula. In 1420, the Portuguese planted the first stalks of sugar cane on Madeira Island. From there it spread to West Africa, the Azores, and the Canary Islands. During a stop at the Canaries in 1493, Christopher Columbus collected sugar cane, which he planted on the island of Santo Domingo.

In addition to sweetening foods, sugar was also thought to have strong medicinal qualities. Practitioners of Ayurvedic medicine prescribed sugar for bronchitis, skin rashes, urinary tract infections, and constipation. An Asian folk remedy suggests mixing sugar with ginger to cure hiccups.

Today, approximately 100 of the world's 180 countries produce sugar, but all countries consume it. In Ethiopia, which has one of the lowest rates of sugar consumption in the world, each person still eats 6.6 pounds per year. Belgium, at 88 pounds per person per year, has one of the highest rates (the delicious Belgian chocolates probably account for the high level of consumption).

While sugar cane is packed with vitamins and minerals, refined sugar is stripped of almost all nutrients. Furthermore, refined sugar in a variety of foods—from bread to juices and snacks—is blamed for rising rates of obesity, diabetes, and, of course, tooth decay.

1000

B.C.

A Caribbean sugarcane plantation

CLOCK

The sun acted as the first clock. About 7,000 or 8,000 years ago, people in the Near East and northern Africa began to pay attention to the time of day, most likely as a result of religious and political concerns—that the king showed himself at a certain time or that sacrifices were offered to the gods at an auspicious hour. The Egyptians used their tall obelisks as monument sundials—the shadow cast by the pointed stone indicated the time of the day. There was no way to mark the hours after dark, of course.

By definition, a *clock* is a mechanism or device that by a constant repetitive action marks the passage of time. By 1500 B.C., the Egyptians had invented a true clock that used water to measure the hours. Water dripped at a regular rate from a hole in the bottom of a stone vessel. To tell the time, a person looked inside the jar at the markings scored on the inside walls; the water level indicated what time it was.

The mechanical clocks that appeared in Europe in the fourteenth century, usually set in church towers, marked the hours but not the minutes. The word *clock*, by the way, comes from the French word *cloche*, meaning bell, which makes perfect sense since the method of timekeeping in medieval villages and towns was the regular peal of the church bell from the steeple—the same place where the newfangled clock was installed. The minute hand of a clock was introduced in 1577 by a Swiss clockmaker, Jost Burgi (1552–1632), at the request of the astronomer Tycho Brahe, who needed an accurate timepiece to chart the movement of celestial bodies.

All of these clocks were very large, too big to be brought into a home. One of the first to mass produce moderately priced shelf clocks was a Connecticut Yankee, Eli Terry (1772–1852). He made interchangeable clock components out of wood or brass (the brass clocks were more expensive); in 1808 alone he had orders for 4,000 clocks.

One of the workers in Terry's factory went on to start his own clock business—Seth Thomas, whose name is still synonymous with quality timepieces.

The inner workings of Big Ben clocktower, London, England

Loom

To understand weaving, one must learn two important terms: warp and weft. The *warp* are the threads that are stretched taut on a loom. The *weft* are the threads wound onto a wooden handle known as a shuttle and passed back and forth through the warp. For more than 7,000 years, that is how cloth has been made.

The earliest looms were tree branches. The weaver tied his or her warp threads to the limb of a tree, pulled them taut and tied them off on stakes in the ground. This became the model for the frame loom, a wooden box on which the warp threads were strung. Virtually every society on the planet learned how to weave cloth, and various tribal groups in Asia developed the high art of weaving carpets.

For thousands of years, the loom, like weaving itself, changed hardly at all. Weaving was something all homemakers did for their families; professional weavers tended to work out of their homes. The first textile factory did not come along until 1785 when Edmund Cartwright (1743–1823), an Anglican clergyman from Nottinghamshire in England, invented the first mechanical loom. With his brother Major John Cartwright he built the first textile mill in Manchester, but a year after it opened it burned to the ground—most likely a target of arsonists. Rev. Cartwright's mechanical loom drove the professional hand weavers out of business, sparking a violent reaction among working men and women who feared machines would take away their livelihood. Mobs of such anti-mechanization laborers known as Luddites attacked mills and factories, breaking up machinery such as the Cartwrights' looms. Of course, there was no stopping the mechanized looms—they were fast, efficient, and produced cloth cheaply.

Interestingly, as part of his independence movement, Mahatma Gandhi urged his followers to stop buying manufactured cloth from Great Britain and return to weaving their own.

5000 B.C.

A man uses a loom to weave a carpet

SAUNA

Saunas are closely associated with Finland, a nation with 5.1 million inhabitants and 1.7 million saunas. The basic procedure involves sitting on wooden benches in a little wooden building, pouring water over hot stones to create steam, traditionally while beating yourself gently with the leafy branches from a birch tree. The word for the experience—*löyly*—is a term from the ancient Finno-Ugric language that dates back about 7,000 years. From Finland, the sauna spread throughout Scandinavia, the Baltic region, and even into Russia, but many other societies have saunas or steam baths, too.

American Indian tribes have a dome-shaped sweat lodge, which they used for ritual purification, and therefore is considered sacred. The ancient Roman bath included a steamy room known as a *calderium*, intended to open the pores to sweat out any dirt. In Japan, sauna-style rooms are available at a *sento*, or public bath, but bathing in natural hot springs is also popular. Sweat lodge–type baths are also found among African tribes, although the temperature tends to be lower (no doubt to compensate for the region's hot climate).

In a traditional Finnish sauna, the occupants are naked, but outside of Finland most users of a sauna are more comfortable wearing at least a towel. Japan has the most elaborate code for saunas and public baths, and one should be considerate and learn the ground rules before entering a Japanese bathhouse. One place where saunas have not gained much headway is the United States. Although they can be found in most health clubs, saunas are rarely used.

Traditionally saunas are considered therapeutic, releasing toxins from the pores, relaxing the mind, muscles, and nerves, and even improving such chronic conditions as asthma and arthritis.

Finnish soldiers indulge in their country's invention, the sauna

Wheel

An ancient Chinese ceremonial chariot with wooden wheels

The wheel was probably developed in Sumer, in ancient Mesopotamia (present-day Iraq) about 6,000 years ago. There is no documentary record that explains how the wheel came to be, but it is likely that our ancestors noticed that it was easier to move a heavy object if it were rolled over something round—a log, for example. However, finding enough fallen logs or collecting them and lining them the entire distance that they would be needed to transport something hefty could not have been easy. A sled or sleigh with runners that could be dragged by human or animal power probably preceded the wheel's development—it's possible that dragging a sled over rollers was the next step. From this point, we arrive at the invention of the wheel mounted on an axle.

This is all speculative, of course, but it is a likely scenario.

From Sumer, the wheel spread throughout the Middle East, into Egypt, and as far as India. It may have been invented independently in Europe and in East Asia—as is the case with the most ancient technological innovations, there is little or no precise evidence. Equally mysterious is why none of the people of the Americas had the wheel. Not even the Olmecs, the Maya, the Aztecs, or the Incas—in spite of the great achievements of their civilizations—developed the wheel. And here's another intriguing point: If the arrival of true human beings dates to 50,000 years ago, it took our ancestors 44,000 years to figure out they didn't have to walk all the time.

Of course, for optimum performance, wheeled vehicles require roads. Persia built a road 1,600-miles long in the sixth century B.C. The ancient Romans constructed a network of roads that covered 50,000 miles. By the third century A.D., the Chinese had laid down nine-lane highways, with the center lane reserved for the exclusive use of the imperial family.

4000 B.C.

Third century B.C. ivory mosaic of a horse-drawn chariot

Wigs

In ancient Egypt, everyone—men, women, and children—shaved their heads. Being bald served two purposes: first, it eliminated the possibility of becoming infested with lice and, second, in the scorching temperatures of an Egyptian summer it was much more comfortable to be without hair. That said, the Egyptians, women especially, wore wigs when they received visitors in their homes or when they left the house. The wig was a symbol of status and had the practical advantage of shielding the wearer's head from the sun.

Women of ancient Greece and Rome kept their natural hair but also wore elaborate wigs on formal occasions. To meet the large demand for blond wigs, wigmakers traveled to northern Europe to purchase the hair of blond barbarians or they just cut off the hair of blond-haired slaves they happened to own.

From the fall of the Roman Empire in the fourth and fifth centuries until the sixteenth century, wigs were out of fashion in Europe. The style returned in the late 1500s and one of its most enthusiastic supporters was Queen Elizabeth I of England. In her portraits, she always shown wearing a wig of tightly curled red hair.

After going bald at a young age, King Louis XIII of France began to wear a wig, which started a new fashion among men. In imitation of the French, King Charles II of England introduced to his court wigs that were shoulder length or longer.

Wigs remained popular into the eighteenth century, and then the style was to wear wigs that had been powdered white. The powder was finely ground starch. After applying powder to the wig, it was typically scented with perfume. Since wigs were associated with royalty, the fashion for wigs died out in the United States after the American Revolution and in France after the French Revolution.

Today wearing wigs (or toupees in the case of men), made of human hair or of some synthetic fiber, is a matter of personal taste.

4000 B.C.

The Egyptian mummies were decorated with natural hair wigs

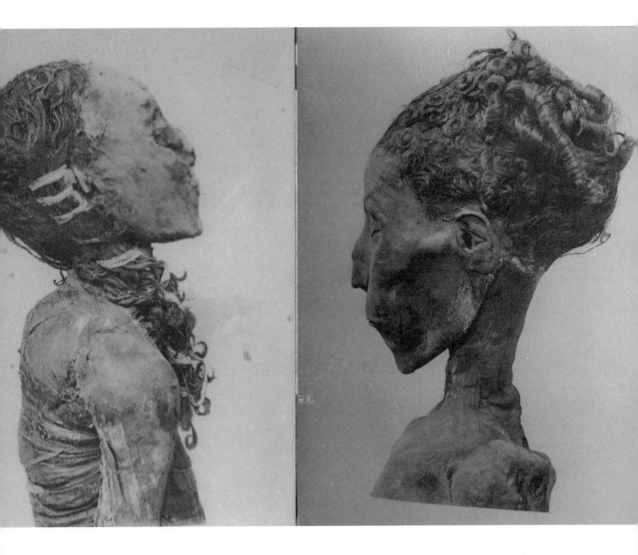

SUNDIAL

The first and certainly the simplest timepiece was a stick stuck upright in the ground; people guesstimated the time of day by the length of the shadow cast by the stick. When there was no shadow, it was high noon and the sun was directly overhead.

One of the earliest surviving sundials was made in Egypt around 800 B.C. It was carved from a block of green stone known as schist; it had an upright piece to cast a shadow. Six marks were carved into the face of the stone, and as the shadow from the upright piece fell on these marks, the Egyptians were able to tell the time of day.

Around 250 B.C., the Greeks, thanks to their love for geometry, began devising more accurate sundials. Apollonius of Perga constructed a sundial that had a line for each hour of the day. About a century later, the architect and engineer Marcus Vitruvius Pollio set up the first public sundial in Rome. In his book, *De architectura* (*Concerning Architecture*), he mentions that portable sundials are available in Rome, which would make them the first pocket watches.

Sundials remained the primary timekeeper well into the Renaissance period. Even after the construction of the first mechanical clocks in Europe in the 1300s, sundials remained popular. By the eighteenth century, when most church steeples had a clock—and mantle clocks, wall clocks, and pocket watches were readily available—sun dials were relegated to the garden as an elegant ornament. Nonetheless, unlike clocks and watches which had to be wound and tended, and their mechanisms adjusted to ensure they kept accurate time, a sundial was regarded as constant and reliable. Hence, in 1777, when the Marquis de Lafayette wanted to express his admiration for his friend and mentor, General George Washington, he presented him, not with a gold watch, but with a silver sundial.

4000 B.C.

Sixteenth-century sundials discovered in the H.M.S. *Mary Rose* shipwreck

PLOW

A horse drawn "beet puller" from 1915

By 7000 B.C., grain, peas, and lentils were under cultivation in the Near East, rice in China, taro in New Guinea, and beans and squash in South America (soon to be followed by the potato). To loosen the soil and make it easier to plant crops, the first farmers used a primitive plow called an ard. It was a beam with a sharpened stick at one end and an ox hitched to the other. It did not cut into the soil so much as it skidded and bounced along the surface. The Chinese discovered that a sharpened piece of metal would slice into the soil, making a rough furrow.

By the year 1000 farmers in Europe used a sophisticated plow that consisted of a knifelike blade known as a coulter that cut open the turf. Behind the coulter was the triangular plowshare that sliced through the topsoil. The soil rode up a curved piece of wood behind the plowshare, known as the moldboard, falling over into a neat furrow. Because the soil often clung to the plowshare or the moldboard, the farmer often had to stop his team to clean off the blades.

In 1837, an Illinois blacksmith named John Deere (1804–1886) invented a steel plow in which the plowshare and the moldboard were a single piece and the soil did not stick to the blade. In 1839, Deere made ten such plows by hand. In 1841, he received orders for seventy-five of his plows. The next year, he made one hundred plows and realized he had to make the switch from blacksmithing to retailing and manufacturing. In 1849, Deere's factory produced 2,136 plows.

John Deere's improved plow was so effective and easy to use, it enabled small farmers to increase their acreage. Before Deere, existing plows could not cut through the thick turf and dense tall grass of the Great Plains. The steel plow did the job, transforming the plains into fertile farmland.

4000
B.C.

Depiction of farming activities, c. 1250 B.C.

Bronze

The discovery of bronze represents a great leap forward in human civilization, but exactly where or when bronze was first invented is a point that historians and archaeologists continue to debate. The people of the Near East had bronze around 35,000 B.C., but it is possible that the inhabitants of Thailand invented bronze 1,500 years earlier. The debate rages on.

Bronze is made by mixing molten copper with molten tin or molten arsenic. Arsenic, of course, is both a metal and a poison, so blacksmiths who made bronze risked their health and their lives. Since the mythologies of ancient Africa, Scandinavia, Greece, and Rome all tell stories of blacksmithing gods who are lame in some way or another, some scholars believe that the stories grew out of the real-life experiences of blacksmiths. Atrophied muscles is one of the symptoms of arsenic poisoning.

Tin, the other metal used to make bronze, usually was exported from Britain—there were large tin deposits in Cornwall in the southwestern corner of the island. There is evidence of Phoenician traders from North Africa sailing to Cornwall to trade for the tin that they carried back to the markets of the Mediterranean.

Before the discovery of bronze, metal objects were made of copper, a soft metal unsuitable for warfare, and not even particularly good for such things as needles. By mixing copper and tin (the usual proportion is 88 percent copper and 12 percent tin), bronze is created. Bronze is a much stronger metal, stronger even than iron, which made it ideal for weapons, armor, and tools of all kinds. Bronze, when struck, has a musical quality, so the first bells and cymbals were made of bronze.

3500 B.C.

Macedonian sculptures from the Bronze Age

Oven

By 3200 B.C., just about every house in the Indus Valley, in what is now western India, had its own oven for baking bread (it's likely this oven was also used to cook meat and vegetables). The earliest ovens would have been fired with wood and later with charcoal. The earliest types of bread were a paste of coarsely ground grain, water, and perhaps a little salt. Such early forms of bread survive today as Mexican tortillas and Indian chapatis. Just as the ancient Greeks were the first to develop front-loading ovens built into the wall of the house, the Greeks also perfected the art of baking. By adding yeast, fruit, cheese, and honey, the Greeks turned out a steady stream of recipes for breads, cakes, and pastries. According to food historians, by 300 A.D., the Greeks had seventy different recipes for breads of different sized loaves that used a wide variety of ingredients.

Oven design barely changed over the centuries. The interior chamber was built of brick, stone, clay, or even cement—in other words, anything fireproof. It was semicircular in shape with a smooth, flat bottom. A wood fire was built inside and once the interior was hot enough to begin baking or cooking, the fire was moved back, or banked, to the rear or the sides of the oven. Such ovens date back to ancient Greece and Rome; they were used throughout the Middle Ages and the first colonists in North America built such ovens in their homes. The masonry oven was replaced early in the nineteenth century by the cast-iron kitchen stove that also had an oven component. For more than a century, old-fashioned masonry ovens could only be found in bakeries and pizzerias.

Recently, there has been a revival of wood-burning ovens in fine restaurants, artisanal bakeries, and specialty pizzerias. Such ovens show a real commitment to traditional cuisine since they can take up to five hours to get hot enough to begin making pizza, breads, and baked dishes.

3200

B.C.

Mosaic depicting an ancient oven

ALPHABET

An alphabet is a set of symbols each of which represents a sound. To be an alphabet in the purest sense, it must include letters for both vowels and consonants, which sounds like a statement of the obvious until you recall that the Hebrew and Arabic alphabets, for example, only have letters for consonants.

One of the earliest writing systems was cuneiform, developed by the Sumerians around the year 3000 B.C., in what is now Iraq. It had thousands of symbols, which made it extremely difficult to learn. Then, around 1050 B.C., the Phoenicians (who lived in what is now Israel, Lebanon, and Syria) developed an all-consonant alphabet comprised of only two dozen letters. Since it was much easier to learn than cuneiform, it thereby sparked a little boom in literacy in the ancient Middle East. The Phoenicians were a nation of traders who visited markets throughout the Mediterranean and wherever they did business, they brought their alphabet with them; archaeologists have uncovered Phoenician inscriptions in places as wide ranging as Greece, North Africa, and Ireland.

The Greeks adopted the Phoenician alphabet and adapted it by adding symbols for vowels. With symbols for consonants and vowels, the ancient Greek alphabet is the first true alphabet. In the seventh century B.C., the Romans, inspired by the Greeks, developed their own alphabet. The Roman or Latin alphabet had twenty-one letters—the G, J, U, W, and Y were added centuries later. Thanks to the Roman Empire, the Latin alphabet became known throughout the ancient world, from Britain to Persia. Today, it is the most widely used alphabet on the planet.

3000

B.C.

Cuneiform signs were the earliest form of alphabet

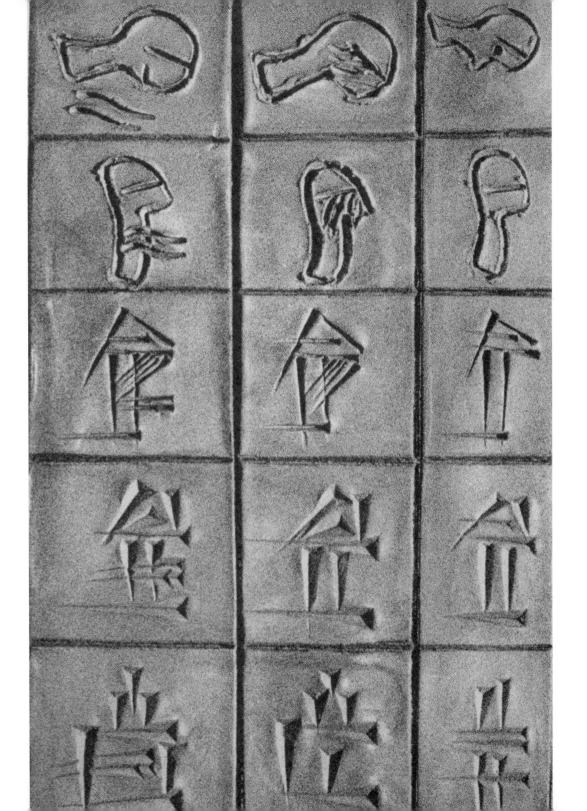

Buttons

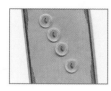

Nearly 5,000 years ago, inhabitants of the Indus Valley in India wore buttons—but as decorative items, not to fasten their clothes. Archaeologists have found buttons from ancient China to ancient Rome and in those places, too, the buttons were worn like a brooch. The clothes fastener of choice in ancient times was the pin.

Buttons that kept clothes closed began to appear in Europe in the 1200s among the nobility. By the 1400s, buttons were in use universally by the middle and the upper classes; the catalyst was a new fashion for form-fitting clothing. At the same time that people began to use buttons, they also began to use laces, which could create an even tighter silhouette.

Today, most buttons are made of plastic or inexpensive metals but in times past, buttons were fashioned from shell, ivory, bone, and wood. Royal portraits from the late Middle Ages and Renaissance era often show monarchs sporting buttons made of gold, silver, or even jewels.

The most common type of buttons include shank buttons, which have a tiny raised ring on the back; covered buttons, which are wrapped in fabric; flat buttons, which are the most common today; and Mandarin buttons, which are elaborate knots of cord (actually, the cord knots aren't called buttons, they are known as frogs).

3000

B.C.

These gold-and-ivory buttons were reportedly produced by Peter the Great

CANDLE

Candles were burning in Egypt and Crete 5,000 years ago. They were made from beeswax or tallow, which is derived from the fat of cattle and sheep. Until the nineteenth century, candles and oil lamps were the only source of light available. Over the centuries, the manufacture of candles has remained virtually unchanged: it is a column of wax or tallow—or in modern times, paraffin—with a wick running up the center of the entire length of the candle. With the development of kerosene in the nineteenth century, most homes and businesses switched to kerosene lamps but they still kept candles in the house for their aesthetic value—the soft warm glow of candlelight remains popular.

Virtually every major religion employs candles to decorate altars and shrines because the desire to keep a light burning near something we love, revere, or respect is rooted deep in the human psyche. To light a candle at a holy place before a sacred image is a sign of religious devotion or is a token sacrifice when asking for heavenly aid. The sacrificial part of the custom is the offering we leave in the coin box—depending on the size of the candle, it can range from a few cents to a couple of dollars.

In the United States, church candles tend to come in two basic sizes: small round candles known today as tea lights, or tall, thick candles usually housed in a glass or plastic container. Europeans bring a bit more flair to the tradition, offering tall, thin, white tapers, in sharp contrast to the stumpy candles available in American churches. During the Middle Ages, it was a common practice among the faithful to promise that if their prayers were answered, they would offer in thanksgiving a candle as heavy and as tall as themselves.

3000 B.C.

Candles are integral to many religious rites, including Communion

Charcoal

An ancient charcoal kiln located in Death Valley National Park

The only times we are likely to encounter charcoal today are as an artist's drawing pencil or as the briquettes stacked in a barbecue grill. In earlier times, however, charcoal was indispensable in moving our ancestors out of the Stone Age and into the age of metal.

It is almost impossible to smelt metal using plain wood. The moisture in the wood combined with other impurities makes it impossible for the fire to get hot enough to perform certain tasks, like melt and shape iron, for example. At least 5,000 years ago, our ancestors discovered that if they burned hardwood slowly the residue, charcoal, would burn at a very high temperature—1000 degrees Celsius, 1832 degrees Fahrenheit—and with very little smoke. Such a high temperature was precisely what was needed to move humankind into the Iron Age. Black, sooty charcoal, then, was an essential tool of modern technology and a building block of civilization.

The classic method for making charcoal is to build a type of kiln. First, a log is set up and the area immediately around it is left open—this will be the kiln's chimney. Then wood is piled in a semicircular mound on either side of the log. The whole pile is covered with earth or turf, although a few small air holes are left open at the bottom. Kindling is thrown down into the chimney area, followed by a burning torch. Once the kindling had caught fire the opening was covered so the intense heat of the kiln would not escape. By managing the opening and sealing of the air holes at the base of the kiln, the charcoal makers ensured that the fuel smoldered without ever bursting into flame (a flare-up would reduce the wood to ash rather than charcoal). This ten-day-long process required careful attention.

Charcoal dominated the metallurgy market until 1709, when an English ironworker, Abraham Darby, discovered how to make coke from coal. Since coke burns twice as hot as charcoal, it became the new preferred fuel in the metal industry and charcoal was relegated to the backyard grill.

Charcoal is traditionally made by the slow burning of hardwood

INK

The first formula for ink was perfected in China: soot from burned pine logs mixed with lamp oil, gelatin from asses' skins, and a couple of drops of musk to mask the foul smell of the lamp oil. It's interesting that ink was invented at about the same time as paper.

Around 400 A.D., a new type of ink came into use made from iron salts (created by applying sulfuric acid to a piece of iron), tannins from nutgalls of an oak tree, and a resin as a thickener. The Latin term for this ink was *encaustum*, which means "to burn in," because this type of ink penetrated the pores of the parchment or papyrus as if it had been burned in. Initially, this ink appeared bluish-black on the page, then over time darkened to deep black, before fading to brown with the passage of the centuries.

When Johannes Gutenberg invented the printing press, the iron salts, nutgall, and resin ink proved to be unsuitable: It didn't adhere well to the metal type and blurred on the page. A new type of ink was developed specifically for the printing press using soot, turpentine, and walnut oil. This formula clung to the type and printed clean, sharp characters on the page, giving the letters an attractive, shiny appearance.

Traditionally, invisible inks have not been inks at all. Milk, lemon, orange, apple, or onion juice, even urine, have been used to write secret messages. The invisible message becomes visible when it is held up to a source of heat, such as a candle flame.

3000

B.C.

An eleventh-century Chinese civil service exam

Paper

Modern paper is produced in oversized rolls before being cut to size

Before there was paper, there was clay. The scribes, poets, and accountants of Mesopotamia wrote on tablets of wet clay that were baked in a kiln. What emerged from the oven was a hard flat piece of pottery that was so durable archaeologists have uncovered thousands upon thousands of them, most of them still in excellent condition. Aside from durability, however, baked clay tablets were not practical: They took up a lot of space; they were heavy; and once the tablet had been baked, there was no correcting, editing, or adding to the document.

The solution to the sturdy but ungainly clay tablet was a sheet of papyrus. In its natural state, papyrus is a reed that grows in the marshes of the Nile. The ancient Egyptians discovered that the pith, the soft white tissue inside the stalk, could be pounded into a flat strip of what we would call paper. Light, compact, and portable, the advantages of papyrus were obvious from the beginning. Very quickly, the creation of papyrus scrolls for sale in overseas markets became a boom industry in Egypt.

About the year 105 A.D., the Chinese began making paper from mulberry leaves, linen rags, and bits of hemp. The ingredients were reduced to a pulp in a tub of water and then lifted out with a fine sieve, which allowed the water to drain away. Once the sheet of pulp paper had dried, it was trimmed to the required size and the process was repeated. By the eighth century, the process had spread to Arab lands; during the Crusades, Arabs taught the technique to the Europeans. In Europe at this time, almost all books and documents were on parchment, thin dried sheets of calf, sheep, or goatskin. Making parchment was labor intensive, and it was expensive. Paper, on the other hand, was easy to make and cheap.

Today, most paper is made from wood pulp, which has presented a fresh problem: Paper from wood pulp is highly acidic and disintegrates over time. Many papermakers, therefore, have started producing acid-free paper.

3000 B.C.

A papyrus sheet from the second century B.C.

Sewer System

When we think of great civilizations, we think of ancient Sumer, Egypt, Greece, and Rome. But none of these places developed the first true sewer system—that honor belongs to the village of Skara Brae in Scotland's Orkney Islands. Five thousand years ago, the islanders built small toilet chambers in their houses. The commodes stood over a pit that flowed into an underground stone drain 14 inches wide by 24 inches high. Probably assisted by a bucket of water poured down the commode hole, the human waste flowed down the drain, beyond the boundaries of the village to a cliff, where it was flushed into the sea.

A couple of centuries later, city planners in India and Pakistan developed an elaborate network of drains that ran just below the surface of the street (which made it easy to reach any blockage in the sewer system). About 600 B.C., the Etruscan king of Rome, Lucius Tarquinus Priscus (reigned 616 B.C. –579 B.C.) constructed a sewer so large you could drive a cart through it. The Cloaca Maxima, the Great Sewer, was so impressive, the Romans even gave the sewer its own goddess, Cloacina. The sewer ran through the middle of Rome, collecting all the nastiness of the city into one large channel. Incredibly, for 600 years, it was an open drain; in 33 B.C., Caesar Augustus had the thing covered over to eliminate the stench.

In addition to collecting human waste, runoff from the streets, and water from the public baths, the Cloaca Maxima was also a dumping ground for bodies. The assassins of Emperor Elagabalus (203–222) disposed of his corpse in the Cloaca Maxima, and the Emperor Diocletian, after executing St. Sebastian (died 287), ordered the martyr's body dumped in the sewer.

3000

B.C.

A Viennese sewer cleaner

SILK

Silkworms no longer exist in the wild; after 5,000 years of nonstop domestication, they are entirely dependent on humans for everything, from their food to their reproduction. Silk is the filament the silkworm spins out in one continuous shimmering thread when it makes its cocoon. The silkworm, which is actually a larva, makes a cocoon in preparation for its transformation into a moth—this facet of the silkworm's life is controlled, too. A silk moth would probably damage the silk cocoon, so most silkworms are not permitted to transform.

Five thousand years ago, the Chinese were already domesticating the silkworm, collecting the leaves of the white mulberry tree to feed the creatures, then killing them by steaming the finished cocoon or leaving it in direct sunlight. Once the silkworm was dead, the filament was unwound and spun into threads of silk. The Chinese kept the secret of silk entirely to themselves for about 3,000 years but then it was leaked to Korea and India. In 552, Christian monks smuggled a few silkworms out of China by concealing them inside their hollowed-out walking staffs. They delivered the silkworms to Emperor Justinian in Constantinople, who immediately made silk an imperial monopoly, setting up a silk factory within the palace. In spite of the efforts of the Chinese and the Byzantines, the secret of silk-making spread across Europe and the Arab world, yet it still remained a luxury item.

Silk's shimmering quality is the result of the distinctive triangular shape of its fibers, which refract light at different angles, so that light appears to move in waves across the fabric.

Early in the seventeenth century, King James I tried to introduce the silk industry to the colony of Virginia but it made no headway. Shaker communities in Kentucky did produce silk in the nineteenth century. In the late 1800s, silk factories opened in Paterson, New Jersey, although after a silk workers' strike in 1913 most of these closed and moved to the South where labor unions were not so active. Today, China and Japan are the world's leading producers of silk.

The Chinese were the first to domesticate the silkworm

VENEER

Poorly made, mass-produced furniture has given veneer a bad reputation in recent years, yet for millennia veneer was regarded as an art form. For example, in about the year 50 B.C., Cleopatra presented her lover, Julius Caesar, with a beautiful table decorated with veneers of fine wood and decorative ivory inlay.

It is possible that the art of veneer was invented in Egypt (the oldest surviving example was found in the tomb of the pharaoh Semerkhet, who reigned about 2950 B.C.). The purpose of veneer is to maximize the use of fine or rare woods, such as ebony. One ebony log might produce one table, but by using thin sheets of ebony a master craftsman could embellish many tables, chests, and other pieces of furniture. Even so, because the only tool the Egyptians had for such work was the saw, their thinnest veneers were about one-quarter of an inch thick.

The heyday of veneers began in seventeenth-century Europe. At the time, metalworkers were beginning to produce fine cutting instruments made of steel rather than iron. Many of these blades were made as surgical equipment but cabinetmakers purchased them, too, as they enabled them to turn out much finer work. In the eighteenth century, Thomas Chippendale, George Hepplewhite, and Thomas Sheraton—three legendary masters of the art of making fine furniture—were handcrafting veneers of satinwood, tulipwood, and rosewood, with meticulous inlays depicting fans, shells, and classical motifs. Chippendale and his colleagues still applied their veneers as the ancient Egyptians had, gluing the sheets to the underlying wood, but their veneers were finer, about one-eighth of an inch thick.

The type of veneer we are most likely to encounter today is, interestingly, plywood. In this case, the wood is not very fine but sheets of plywood have the advantage of being sturdy and inexpensive. And, just like fine veneers, the sheets of spruce, fir, or pine are glued together.

An eighteenth-century commode decorated with ebony veneer and ivory inlay

UMBRELLA

The word *umbrella* comes from the Latin word *umbra,* which means shade or shadow. The earliest known umbrella was made in Akkadia, modern-day Iraq, during the reign of King Sargon (ca. 2334 B.C. –2279 B.C.). A carving of Sargon on a stone victory monument shows him leading his troops into battle as an attendant holds a large umbrella over him. In the years that follow, we find depictions of umbrellas in Cyprus, Greece, and Egypt, but all of these umbrellas were used as sun shades: The first waterproof umbrella to protect the carrier from the rain was invented in China somewhere between the years 386 and 535 A.D. The Chinese umbrella was made of heavy mulberry paper, sealed with oil to make it rainproof. The emperor was protected by a red-and-yellow umbrella; everyone else in Chinese society carried blue umbrellas.

The first umbrella in Europe was introduced by Giovanni de Marignolli (ca. 1290 to ca. 1360), a Franciscan priest from Florence, who was sent as the pope's emissary to the emperor of China. While traveling through India, Father Giovanni saw people carrying umbrellas and he brought one back to Florence. It didn't catch on.

The man who popularized the umbrella in Europe was the English travel writer Jonas Hanway (1712–1786); he purchased umbrellas while in Persia and used them back in London for the next thirty years. The first umbrella shop, James Smith & Sons, opened in London in 1830. The frames of nineteenth-century umbrellas were made of wood or whalebone, the fabric was oiled cloth or alpaca, and the handles were carved ebony or some other hardwood.

Little paper cocktail umbrellas became a trademark in the 1930s at Victor Bergeron's San Francisco bar-and-restaurant, Trader Vic's. Bergeron admitted, however, that he took the idea from Donn Gantt, founder of the Don the Beachcomber chain of Polynesian-style restaurants. Originally, Gantt got the idea from Chinese restaurants of the 1910s and 1920s.

2400 B.C.

Sudden Shower on Ohashi Bridge at Ataka

BELLS

We tend to thinks of bells as a musical, if old-fashioned, device for marking the hours. Yet the earliest bells were made for different purposes.

The oldest bells were made on the island of Crete about 4,000 years ago. They were small, made of clay, and inside hung a wooden clapper. These bells were hung in trees where their sound, it was believed, would attract benevolent spirits.

The first metal bells, made of bronze, were probably cast by the Chinese about 3,500 years ago. The Chinese bells did not have a clapper; to sound them they were struck with a mallet. Bell makers discovered that a bell could give off two distinct notes—one if struck at the center, another if struck on its edge. In 1978, archaeologists discovered a set of sixty-four bells in the fifth-century B.C. tomb of the marchioness of Yi—each with perfect pitch. The bells were played like a set of chimes or a xylophone.

In medieval Europe, bell makers also learned that bells of different sizes produced different tones. Large churches and cathedrals began to commission many bells for their towers where trained bell ringers produced glorious sounds by "playing" the bells. Ringing such a set of bells is called "change ringing"; it's a dying art, rarely heard in the United States, where almost all churches have given up ringing their bells by hand in favor of an electronic chimes system.

As the number of bells in a steeple increased, there arose the custom of naming each one. The nineteen bells of St. Patrick's Cathedral in New York City, for example, were all named for various saints at the time of their dedication in 1897. Before the bells were hung, Archbishop Michael Corrigan performed the rare rite of "baptizing" the bells. He anointed each bell with holy oil, washed it with holy water, and then burned incense beneath the bell, filling its hollow with fragrance.

A Korean Buddhist monk strikes a bell

GLASS

The secret to making glass was discovered in the Near East 4,000 years ago. The first glass objects were very small—seals, beads, and bits of mosaic that could be inlaid into jewelry.

To make glass, one needs quartz or silica sand, the type of sand typically found on beaches. When it is superheated, silica sand melts into a smooth liquid, and when this liquid cools is hardens into a transparent substance we call glass. By the first century B.C., the Romans were using panes of glass in their windows. In the Roman city of Pompeii (near present-day Naples), archaeologists have found glass panes set in a bronze frame with two pivots that permitted the homeowner to open and close the window. The largest windows found in Pompeii measured 40″ x 28″; the panes were half an inch thick and had been rubbed with sand on one side giving them the appearance of frosted glass.

The technique of glassblowing was discovered in the first century by the Phoenicians, in what is now Lebanon. Soon, inexpensive glass bowls, cups, and other common household objects were available in marketplaces throughout the Mediterranean world, replacing wooden objects that weren't nearly as pretty and metal objects that were much more expensive.

There is a story—and it may be an ancient Roman urban legend—that a man discovered the secret to making shatterproof glass. He took his new invention to the Emperor Tiberius, expecting that he would receive a fat reward. "Are we two the only ones who know about this?" the emperor asked the inventor. When the man said yes, Tiberius called his guards and had the man executed. Unbreakable glass, the emperor reasoned, would put the entire glassmaking industry out of business.

Superheated glass

RIVET

One of the earliest fasteners on the planet, the rivet has changed scarcely at all in the last 4,000 years. The earliest rivet discovered by archaeologists dates back to the Bronze Age, which makes perfect sense since a rivet must be metal in order to work properly.

Unlike a nail, a rivet does not have a sharp point, and unlike a screw, its shaft is smooth and there is no indentation in the head so it can be turned by a screwdriver. To install a rivet you push it through a pre-drilled hole, apply pressure against the cap end to keep it immobile, then with a hammer or mallet strike the other end until it is flattened. This creates a very tight fit, more secure than any nail or screw.

In hardware parlance, the cap end of the rivet is known as the factory head (because it was made that way at the rivet factory) and the smashed-up end is called the bucktail—for reasons no one can explain.

Although a very old type of fastener, rivets were the joiner of choice for major metal construction projects into the twentieth century. The pieces of the Eiffel Tower and the Sydney Harbour Bridge in Australia were riveted together. Approximately three million rivets were used to hold together the thousands of steel plates of the *Titanic*. And, of course, the Levi Strauss Company uses countless tiny copper rivets to reinforce its blue jeans.

During World War II, Rosie the Riveter became a national icon, a symbol of the can-do spirit of the six million American women who took jobs in factories and manufacturing plants while millions of American men were fighting in Europe and Asia. Supposedly, Rosie was based on a real person—Rose Will Monroe of Kentucky who moved north during the war and took a job as a riveter building B-29 and B-24 bombers at the Willow Run Aircraft Factory in Ypsilanti, Michigan.

Female machinists were a major part of the war effort during WWII

Soap

The story of notions of cleanliness and the importance of bathing is a complex one and varies greatly from society to society. A much simpler story is the origin of soap: The Babylonians discovered that if oil was boiled with an alkali such as potash the result was a soft waxy substance that washed away dirt and grime. This was the first soap.

People in other parts of the ancient world were as concerned as the Babylonians with hygiene: The Egyptians and the Jews washed themselves with sodium carbonate, also know as washing soda, which is still used in some detergents. The Greeks took an exfoliant approach, rubbing their bodies with sand or pumice stone. The Chinese rubbed themselves clean with waxy soap-beans from the saponin tree. Among the Romans, washing was a two-person job: They rubbed their bodies with olive oil then had a relative, friend, or slave scrape the oil and dirt off their body with a hand tool know as a *strigil*, which resembled a small sickle. Thanks to their conquest of Western Europe, Romans had access to a product from Germany known as sapo—balls of goat fat mixed with the ash of beechwood trees. The Germans washed with it but the Romans used it to dye their hair reddish blond.

The Arabs first introduced bars or cakes of soap, such as we buy today, around the ninth century—manufacturing them was a major industry in Moorish Spain. Scented soaps first appeared during the Middle Ages. By 1300, the English were selling heavily perfumed soft soaps packed in wooden bowls.

2000 B.C.

Porcelain

Usually porcelain is called *china*, not only because China is the place where it was invented 3,500 years ago but also because for many centuries all porcelain came from China. Porcelain differed from pottery in several respects—even its most delicate shapes such as teacups were harder and more durable; it had a smooth finish that has often been compared to enamel; it is translucent; and it inspired Chinese artists to experiment with exquisite glazes and patterns such as had never been seen on prosaic clay pots. Porcelain was one of the world's earliest forms of art.

The secret of Chinese porcelain is a combination of pure white clay, known as kaolin, and petuntse, a type of feldspar unique to China. This mixture must be fired at a higher temperature than common pottery—about 1400 degrees Celsius (hotter than 2500 degrees Fahrenheit). The kaolin and petuntse mix is also known as hard paste porcelain, the classic variety of Chinese porcelain. Other forms of porcelain—soft paste and bone china—were developed in Europe in the seventeenth and eighteenth centuries.

The first Chinese porcelain arrived in Europe around the 1100. It was rare and very expensive and remained a luxury item until the early seventeenth century, when international trade organizations such as the Dutch East India Company began bringing home boatloads of Chinese porcelain. Demand was so high that Chinese porcelain makers began producing designs exclusively for the European market.

European soft paste porcelain manufacturers such as Sevres in France and Meissen in Germany, and bone china manufacturers such as Worcester in England, produced lovely work but the demand for true Chinese porcelain remained steady well into the nineteenth century. At that time, many American seafaring families built fortunes in the China trade, carrying vast quantities of fine porcelain to the United States and Europe.

1500

B.C.

Porcelain is made from pure white clay known as kaolin

SCISSORS

Archaeologists have found possibly the earliest scissors in the remains of Emar, an ancient city in northern Syria. They are made from a single piece of metal, which has been bent into a U-shape. The upper arms of the U are sharpened blades; the user operates the scissors by squeezing and relaxing the U. Using them must have been hard work that required strong hands. Very likely, these scissors or shears were used to cut fabric and leather, as well as to shear sheep and goats.

The Egyptians, ever the master artisans of the ancient world, turned ordinary scissors into works of art. One pair of bronze scissors was decorated with a figure of a man on one blade and a figure of a woman on the other. Lovely as the Egyptian scissors may have been, they were still the U-shape version of the scissors from Emar.

Cross-bladed scissors joined by a pivot with holes for the thumb and index finger, such as we use today, were invented by the Romans during the first century A.D. However, the invasions that destroyed the Roman Empire also damaged much of the physical culture and technology of the Roman world. A Spanish bishop at the time, St. Isidore of Seville, compiled a twenty-volume encyclopedia of all existing knowledge so that the achievements of Rome would not be lost. In his encyclopedia, he describes the Roman's ingenious—and much easier to use—cross-blade scissors. Because they required less brute force and, thanks to the holes for the fingers, could be more precisely controlled, St. Isidore tells us that Roman scissors were used for finer work, such as cutting and styling hair.

Pinking shears were invented in 1893 by a Louise Austin of Whatcom, Washington. Austin's shears had two saw-toothed blades, which enabled a tailor or seamstress to make decorative cuts in fabric. Her pinking shears were also very good when cutting woven cloth: an unfinished edge of cloth cut with standard scissors frays and unravels easily; a cut made by pinking shears limits the amount of fraying, keeping the cloth relatively stable until the sewer is ready to hem it.

1400
B.C.

Surgical instruments used to operate on Napoleon I

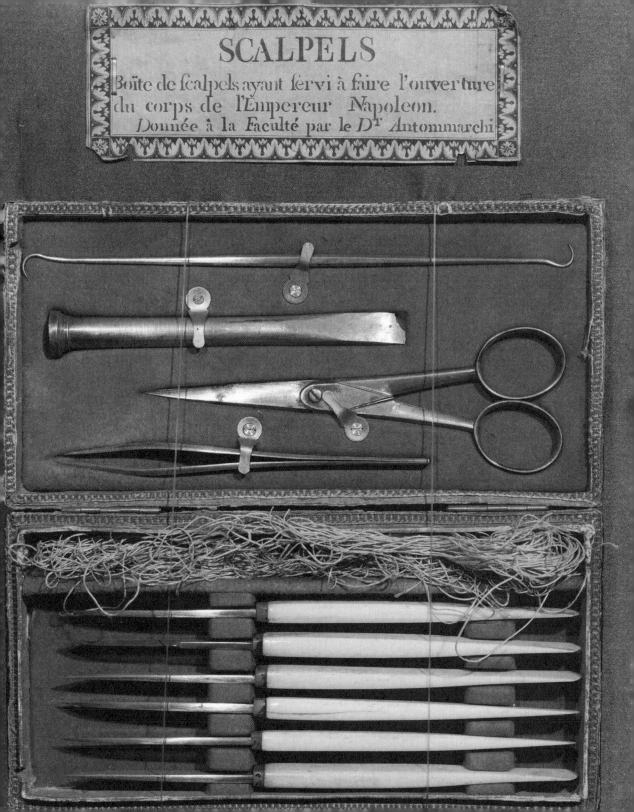

CARTS AND CHARIOTS

A chariot from the Vittalla Temple in India

The first carts got rolling in Mesopotamia about 3,000 years ago. They were heavy vehicles with solid wooden wheels and were pulled by teams of oxen. The technological breakthrough in cart construction was the development of spoked wheels, which were strong but light. They enabled the creation of lighter conveyances such as two-wheeled chariots for use in warfare and for transporting passengers, as well as four-wheeled chariots for transporting merchandise.

In ancient times, chariots could be be found throughout Asia and Europe—from Ireland to China, from Denmark to Egypt. Two, three, or four horses could pull the chariots, although battle chariots tended to be drawn by two horses. For more than a thousand years, the most frightful thing on any battlefield was a mass of chariots bearing down on ranks of infantrymen. It was Alexander the Great who made them obsolete. At the Battle of Gaugamela, the king of Persia, Darius III, ordered his chariots to charge the Greek lines but Alexander had foreseen that. He ordered his men to open their lines and let the chariots pass through. Then Alexander sent his cavalry to attack the chariots from the rear. The strategy was so simple and effective that it became obvious to commanders throughout the Mediterranean world that the days of the chariot as a war machine were over.

Chariots enjoyed a second life at the races. By the eighth century B.C., Greeks were attending chariot races; Homer even describes a race in the *Iliad*, as part of the funeral games for Patroclus. By 680 B.C., chariot racing with four-horse and two-horse teams had been added to the Olympic Games.

The Romans picked up chariot racing from the Greeks and it became such a popular sporting events that the caesars built an enormous track, the Circus Maximus, to showcase the races. The stadium could accommodate 250,000 eager spectators who came as much for the carnage—accidents were a frequent occurrence at chariot races—as for the thrill of the race. A rare exception to this many-accidents rule was the charioteer Scorpus (first century A.D.) who won 2,048 races before finally getting killed in a crash on the track when he was only twenty-seven years old.

Frieze featuring an ancient chariot

FALSE TEETH

Three thousand years ago, north of Rome in the province of Tuscany, lived a sophisticated, creative people known as the Etruscans. Around the year 700 B.C., Etruscan goldsmiths teamed up with Etruscan dentists to make the first dentures either from actual human teeth or false teeth carved from ox teeth. In the case of bridgework or partial dentures, the teeth were set in bands of gold and fitted inside the patient's mouth. At the patient's request, the dentist could either affix the partial permanently among the natural teeth, or could make it removable for easier cleaning. The gold bands, by the way, were carefully positioned to keep them well away from the gums, thereby avoiding discomfort.

By 600 A.D., decorative false teeth had become popular among the Maya in Central America. Archaeologists have discovered a human skull that bears three false teeth made from carved shell. It is possible that the man had three healthy teeth removed for the sake of the shell teeth.

Five thousand years ago, dentists had a thriving practice in Egypt, where so much sand got into the food that it wore off tooth enamel, hastening tooth decay. In most cases, the Egyptian dentists simply yanked out the bad tooth but they did not make replacements—it was considered taboo to interfere with nature.

In 1774, a French pharmacist, Alexis Duchateau, collaborated with a French dentist, Nicholas Dubois De Chemant, to make false teeth of porcelain. Durable and more attractive than teeth made from animal bones, the porcelain teeth were a great success. In 1789, when Dr. De Chemant fled to England to escape the French Revolution, he took his secret for making porcelain teeth with him. From England, porcelain teeth came to the United States where many people, even the renowned portrait artist, Charles Peale, made porcelain dentures. By 1844, the S.S. White Company, founded by Samuel Stockton White of Pennsylvania, made a fortune by mass-producing porcelain false teeth.

700 B.C.

Dental technicians construct steel and plastic dentures

SADDLE

Humans domesticated the horse around 4000 B.C. Very soon thereafter riders probably cinched a blanket over the horse's back to make for a more comfortable ride. The earliest evidence of such a saddle dates from about 700 B.C. in Syria where a rider sat on a pad strapped around the horse's belly. Before long, riders began to embellish their saddles with elaborate embroidery, even gold and precious stones if they were of royal or noble birth. An opulent saddle became a sign of the rider's status.

The solid wooden frame, known as a "tree," was developed late—about 200 A.D. in Rome. Covered with leather and cloth to protect the horse's back from chafing, the tree gave the rider a more stable ride, although at this point the Romans did not have stirrups. The tree also made life better for the horse—by raising the rider off the horse's back it eliminated the pressure point on the horse's spine and distributed the rider's weight along either side of the horse's backbone.

During the Middle Ages, knights desired large war horses, known as chargers, that could support the weight of a man in armor while still able to run at a good speed across a field or in the lists of a tournament. Saddles also became larger and stronger at this time, with a high cantle in the rear and pommel in the front to give the knight a little extra stability, making it less likely that he would be unseated in battle or in a joust.

Since the early eighteenth century, saddles have evolved into two basic types: English and Western. The English style was actually designed by a French riding instructor, François Robichon de la Gueriniere (1688–1751), who invented a new, smaller, lighter saddle suitable for a style of horse training known as classical dressage. English fox hunters adopted Gueriniere's saddle, so it became known as an English saddle. The Western saddle is larger and heavier than the English and is designed for long hours of riding. Its origins go back to the Spanish who introduced it to the New World. The basic design was adapted over time by the vaqueros of Mexico and the cowboys of the American West, all of whom worked almost entirely on horseback.

Saddles were a vital part of a knight's equipment

Coins

An early Roman coin

What with debit cards, credit cards, and old-fashioned paper money, coins seem like a nuisance. Yet for ancient economies they were a leap forward. Under the old barter system, if you wanted to go shopping for a new cooking pot, you had to bring a few goats along with you to the marketplace. Coins were an immense improvement because they were portable and, unlike goats, they wouldn't run off.

The first coins date from about 650 B.C. and were invented in Lydia, a kingdom in what is now western Turkey. But even before people made purchases with actual metal discs, there were small items they used in the same manner as coins. Four thousand years ago, shoppers in Egypt and Mesopotamia went to the market carrying strings of bronze rings, which they gave in exchange for goods. By 1500 B.C., the Chinese walked around with cowrie shells jingling in their pockets.

The Lydian coins were shaped like beans and made of a mix of gold and silver. On one side was stamped a lion, the emblem of King Gyges, and on the other side were symbols that indicated the weight of the coin and guaranteed the purity of the gold and silver used to make it. The advantages of coins were so obvious that almost overnight mints sprung up all over the Near East—from the city-states of Greece to the Persian Empire. At about the same time in India, little bars and discs of silver were being used in trade.

The next stage in the history of coinage is the emergence of the counterfeiter. The oldest counterfeit coin discovered so far came from the Greek island of Aegina and dates from about 550 B.C. It looks like a silver coin but its core is copper.

These days most people disparage coins as "loose change." But the United States Treasury, in an effort to cut down on the cost of printing $1 bills (the denomination that wears out the fastest), is trying to persuade the buying public to use new silver dollars. So far, Americans aren't buying it.

An illustration of a worker minting coins in the late nineteenth-century

Aqueduct

An aqueduct in Pont-du-Gard, Provence, France

The word *aqueduct* comes from two Latin words: *aqua*, meaning "water," and *ducere*, meaning "to lead." Although aqueducts are most closely associated with the Romans, other civilizations devised methods for bringing water long distances from where it was plentiful to where it was needed. In the seventh century B.C., the Assyrians built a 50-mile-long aqueduct to carry water to their capital city, Nineveh. In the fifteenth century, the Aztecs had built two aqueducts, each about 12 miles long, to bring fresh water from the Chapultepec springs to Tenochtitlán, now Mexico City.

The Romans built dozens of aqueducts throughout their empire. The city of Rome was fed by eleven aqueducts, which delivered 300 million gallons of fresh water to the city each day. The method of getting the water to the city was both simple and ingenious—the Romans relied entirely on gravity. The channels within the aqueduct were pitched at a very gentle gradient or incline, enough to keep the water moving but not so steep that it would pick up speed along its way and burst out at the end of its journey like a tidal wave. The tiers of arches that are the most striking feature of an aqueduct were built so that each one is just a bit lower than the one before enabling the water to move in a constant stream downhill.

Work crews were assigned to provide regular maintenance to the aqueducts—cleaning channels of leaves, twigs, and other debris, repairing any breaks in the line, and scraping away the buildup of calcium carbonate, the white powdery residue of hard water.

After the collapse of the Roman Empire in the fifth century, the aqueducts fell into disrepair. As the flow of fresh water into the city of Rome dried up, the inhabitants moved out. During the glory days of the empire, Rome had been home to more than one million people; at its lowest point during the Middle Ages, the city's population dropped to 30,000.

Mural of Tenochtitlán with its famous aqueduct

ARCH

The earliest method of construction, exemplified by Stonehenge and the Parthenon, followed a basic design: upright posts or columns with a flat beam or slab laid across the top. Due to the weight of the top beam, the supporting columns must be close together. Spread them apart and the beam will snap and the building will collapse. An arch, however, permits wider openings because the weight of the bricks or stones used in its construction does not press straight down as in a top slab but is displaced along the arch's curve.

It's often said that the Romans invented the arch. In fact, the ancient Persians, Egyptians, and Babylonians all used arches but generally for underground projects such as drains. The Romans were among the first to bring the arch into the light of day.

The Roman arch is also known as a semicircular arch. It is always built with an odd number of curved, wedge-shaped bricks or stones known as *voussoirs*. Dead center at the peak of the arch is the keystone that locks all of the voussoirs together. That said, exactly how the Romans pieced their arches together is a mystery since the voussoirs, without the anchor of the keystone, would topple over. Most likely, the Romans constructed a wooden frame the exact size of the desired arch upon which they laid the bricks or stones. Once the mortar had set the wooden frame was removed.

The semicircular arch with its wide, perfectly rounded shape looks very handsome; the Romans used it for their aqueducts, amphitheaters, palaces, and of course triumphal arches. The trouble with the semicircular arch is that, because of what is know as the force of compression, the arch wants to bulge out at the sides. To counteract this tendency, the Romans built their arches within walls where the weight of the surrounding masonry would keep the arch stable.

During the Middle Ages, architects developed the pointed or Gothic arch. Since these tall, pointed arches deflect the force of compression more efficiently than the classic Roman arch, a Gothic arch is more structurally sound than a semicircular arch.

A Picturesque View of the Capitol in Rome

STIRRUPS

A stirrup is such a simple thing—a curved hoop of metal with a flat bottom that hangs by a sturdy leather strap from either side of a saddle—that it is remarkable that it was not invented until 4,000 years after humans domesticated the horse. The saddle gives the rider a secure mount and the reins enable the rider to control the horse but, without stirrups, the rider has limited maneuverability or even stability on horseback. If the horse bolts or makes an unexpected turn, there is a good chance the rider will be thrown.

Exactly who invented the stirrup and when it was introduced remains a source of ongoing debate but there is reason to believe that the Huns had stirrups. The Huns were legendary archers but, in the ancient world, archers almost always fought on foot. The Huns, however, almost never fought on foot. When they went into battle, they rode as a massive swarm that, at first, looked wild and undisciplined; then, as the Huns drew near their enemy, they broke ranks and rode in elaborate patterns that encircled the enemy, or harried their flanks, or even rode straight into them. As they rode, they fired arrow after arrow from their small, powerful bows. All of the ancient sources agree that the Huns could even turn in their saddles and shoot backwards—a maneuver that would be possible only if the Huns had stirrups.

However, no stirrups have ever been found among Hun graves. Archaeological excavations of Hun graves have turned up saddles but no stirrups. If Hunnish stirrups were just loops of leather, or a leather loop with a small piece of wood at the bottom, they would have decomposed easily in a grave.

Wherever they came from, stirrups were in use in Scandinavia by the sixth century A.D. and in France by the eighth century. Mounted knights during the Middle Ages found the stirrup especially useful—not for archery but for wielding swords, maces, and battle-axes. The stirrup gave the mounted warrior the stability he needed so he could make a really powerful downward swing against an opponent.

Stirrups allowed horseriders to turn in their saddle

Aspirin

Technically, Hippocrates did not pass out little white tablets to his patients suffering from headaches and fevers; instead, he dispensed a tea made from the bark and leaves of the willow tree. But the effect was the same—the willow tea relieved headaches and reduced fevers. The willow tree contains salicin, which as the name suggests is very similar to acetylsalicylic acid, popularly known as aspirin.

In the 1800s, chemists, trying to improve on the work of Hippocrates, derived salicylic acid from willow bark for medical use. It was a complete failure. The chemists' acid was so intense it irritated the patients' stomachs and even the inside of their mouths.

In the late 1890s, a German chemist, Felix Hoffman (1868–1946), was searching for some medication that would ease his father's rheumatism. He came upon the work of a French chemist, Charles Gergardt, who sixty-five years earlier had tried to make a form of salicylic acid that would not produce nasty side effects but would be stronger than Hippocrates' willow tea. Gergardt's solution was too complex and time-consuming to be practical but the germ of an idea was there. Hoffman improved upon Gergardt's research and found a simple, inexpensive way to manufacture acetylsalicylic acid—aspirin. Hoffman, by the way, worked for Friedrich Bayer and Company, the pharmaceutical manufacturer that still produces aspirin.

Aspirin is widely regarded as an over-the-counter wonder drug. In addition to getting rid of headaches, soothing sore muscles, easing the pain associated with arthritis and rheumatism, an aspirin a day can reduce the risk of heart attack and stroke. Studies are underway to determine if aspirin can also reduce the risk of contracting colon and esophageal cancer.

400
B.C.

An employee at Bayer controls the production of aspirin

CATAPULT

A medieval French catapult

Since the eighth century B.C., there had been Greek colonists on the island of Sicily. However the Carthaginians from North Africa also coveted the island. In 399 B.C., the ruler of the Greek colony of Syracuse, Dionysius the Elder (ca. 432–367 B.C.), was gearing up to drive the Carthaginians out of Sicily once and for all. He assembled the best engineers on the island and put them to work inventing new war machines. The result of Dionysius' research and development program was the catapult.

The catapult created for Dionysius did not hurl rocks, rather, it shot heavy 6-foot-long arrows. A massive bow was mounted on a sturdy three-legged wooden base. Attached to the bow was a long, narrow, grooved wooden frame. The arrow was placed in the groove. A metal claw was snapped over the bowstring. Using winches at the end of the frame, the bowstring was drawn back. When it had reached maximum tension, a touch on the trigger released the arrow, which shot out of the catapult with tremendous strength and speed, skewering enemy troops or smashing large holes in enemy ships. During the Middle Ages, archers began using a small, portable version of Dionysius' catapult; this weapon was known as a crossbow. Like its larger cousin, the crossbow gave an arrow more power and speed than a regular bow operated purely by human muscle: crossbow arrows were the first armor-piercing weapons.

The type of catapult that threw stones rather than fired arrows was invented by another Greek from Syracuse, Archimedes (ca. 287–ca. 212 B.C.). Among his many accomplishments as an inventor, mathematician, and physicist, Archimedes explained the principle of buoyancy and invented the first block-and-tackle pulley for lifting heavy objects. His catapult was capable of flinging a stone weighing 175 pounds a distance of 200 yards, smashing anything it struck. These catapults were the world's first artillery.

399

B.C.

A design for a giant catapult by Leonardo Da Vinci

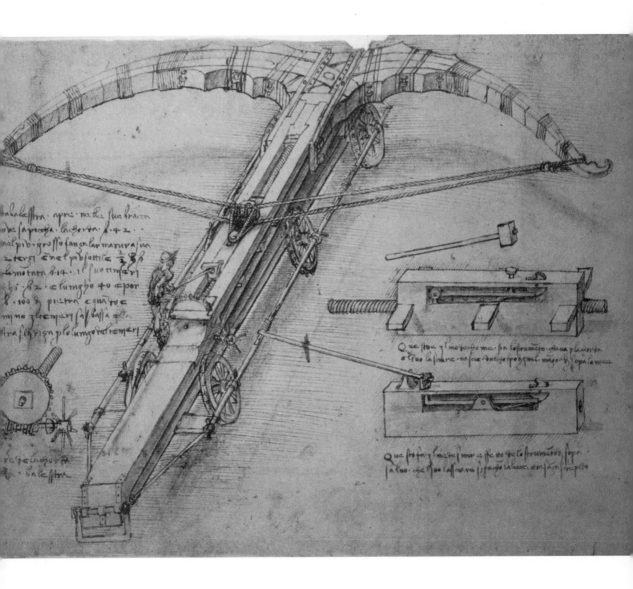

Paved Roads

The Appian Way as it looks today

Before the Goths stormed into the Eternal City on an August day in 410 A.D., the Romans built about 50,000 miles of paved roads throughout their empire, and all of these roads did indeed lead to Rome.

Initially, Roman roads connected the city to places and towns in the vicinity, such as the port of Ostia. The first paved Roman road was the Via Appia, the famous Appian Way, which ran south all the way to Brindisi and the Straits of Messina. Along the Appian Way thousands of slaves who had been captured in the Spartacus rebellion were crucified.

The Romans intended their roads to be permanent. They began by laying a foundation of flagstones on the bare earth. Next followed successive layers of crushed stone and concrete, another layer of flagstone, and then gravel or concrete pounded into the surface to make a hard, smooth pavement. The roads were slightly convex so that rainwater and melting snow would run off to the sides.

The Roman road system not only carried travelers, merchandise, and couriers but the network also was especially efficient at getting the Roman legions from the point where they were encamped to the place where they were needed. In fact, road-building engineers accompanied the legions so that as the army conquered new territory, new roads could be connected to the existing network.

Of course, the Romans were not the only great road builders. In the sixth century B.C., the Persians built the King's Road, which ran 1,600 miles from the capital, Susa, in Iran, to Sardis in Turkey. By 200 A.D., the Chinese had 20,000 miles of roads. And between 1400 and 1520 A.D., the Inca emperors had 15,000 miles of roads constructed throughout their empire in South America. Like the Romans, the Incan road builders were unstoppable—they even threw suspension bridges over river gorges high in the Andes Mountains.

312
B.C.

The Appian Way was the first Roman paved road

LIGHTHOUSE

The world's first lighthouse was such a marvel of its age that it was listed among The Seven Wonders of the World.

After the death of Alexander the Great in 323 B.C., his generals carved up his empire. Ptolemy (367 B.C.–283 B.C.) got Egypt. He abolished the old line of pharaohs, proclaimed himself king under the title Ptolemy I Soter (which means "the savior"), and started a new dynasty that was not Egyptian but Greek. His most famous direct descendant was Cleopatra.

In Ptolemy's day, Alexandria (named for Alexander the Great) was the capital of Egypt. The city had a magnificent harbor but the surrounding countryside was so flat there was no landmark to draw incoming ships to the city. At the entrance to the harbor was a small island, Pharos, which was connected to the shore by a stone mole or causeway about 1,500 feet long. On this island, Ptolemy built his lighthouse. He hired a Greek architect, Sostratos of Cnidus (a city in what is now Turkey), to design and construct the building. Adopting the monumental style of classic Egyptian buildings, Sostratos erected a tower about 450 feet tall. At the summit was a huge polished bronze mirror that would reflect the sun during the day and light from a large beacon fire at night. The light could be seen by ships 35 miles out to sea.

By custom Sostratos was forbidden to sign his masterpiece—only the king's name could appear on it. But Sostratos deceived Ptolemy. He carved into the stone of the lighthouse this inscription, "Sostratos, the son of Dexiphanes, the Cnidian, erected this to the savior gods, on behalf of those who sail the seas." Then he plastered over his inscription and carved into the plaster a new one in honor of Ptolemy. Years later, after the weather had worn away the plaster, it was Sostratos' inscription in stone which all visitors to the lighthouse saw.

The Pharos lighthouse survived until the fourteenth century when two earthquakes in 1303 and 1323 finally reduced this wonder of the world to rubble.

The Great Lighthouse at Alexandria was once considered one of the Seven Wonders of the World

WALLPAPER

The Chinese, who first invented paper, were also the first to paste it on their walls. Their wallpaper was rice paper, which had a subtle texture and was visually very pleasing.

The earliest known decorative wallpaper was created by a French artist, Jean Bourdichon, who painted fifty rolls of paper with a pattern of angels on a blue background. The king of France, Louis XI, took these rolls of wallpaper with him on his travels so that no matter where he was staying, his room would be attractive and familiar. Since this was reusable wallpaper, it was not glued to any surface but simply hung like a tapestry. Inspired by (or perhaps jealous of) the king's rolls of angels, members of the nobility and other kings commissioned artists to paint rolls of paper for them.

The earliest surviving printed wallpaper that was actually glued to a wall has been found at Christ's College in Cambridge, England. It is a pattern of pomegranates dating from 1509. The invention of the printing press accelerated the production of wallpaper in Europe—so much so that, in 1599, a guild of paperhangers was founded in France.

In 1675, French engraver Jean-Michel Papillon designed the first rolls of paper with a continuous pattern that could be matched up on the wall. For this innovation, Papillon is considered the father of modern wallpaper.

Plunket Fleeson printed the first American wallpaper in Philadelphia in 1739. Thrifty Americans used the leftover lengths of wallpaper to line trunks and hatboxes, which is where historians and museum curators have found some of the best-preserved examples of eighteenth-century wallpaper.

In 1839, a printing machine that could produce four-color wallpaper was invented. This was followed by an eight-color printer in 1850 and a twenty-color printer in 1874. Wallpaper became so inexpensive to produce that virtually every budget could afford it. The Victorian era has been called the Golden Age of Wallpaper, with 400 million rolls sold in Europe and the Americas.

The exquisitely wallpapered Salzburg Residenz throne room

CLOTHES IRON

Sometime in the first century B.C., the Chinese began to fill metal pans with hot coals and pass them over fabric to smooth out the wrinkles. By the seventeenth century, Europeans were using cast-iron D-shaped plates, heated in a fire, for the same purpose. They were called sadirons, *sad* being an archaic term for "solid;" and so the act of getting wrinkles out of fabric with this device became known as ironing.

For the next 300 years in Europe and the Americas, the basic shape and construction of a clothes iron changed very little. It was either a flat iron plate heated on a stove or before an open fire, or a cast-iron box into which hot coals were placed. One of the exceptions was Italy, where clothes were smoothed with heated soapstone.

There were irons that were fueled with kerosene or whale oil but these were dangerous—if the iron were accidentally tipped over, it would spill the burning liquid and set the house on fire.

For centuries, ironing was done on the kitchen table or on a length of board laid across two chairs. Usually a thick woolen cloth called an ironing blanket was placed on the table or board so the heat of the iron would not scorch the wood—or worse still, ignite it. In 1858, two Americans, W. Vandenburg, and J. Harvey, patented an ironing board that made it easier to press the sleeves of shirts and blouses as well as pants legs. Vandenburg and Harvey's design has not changed to this day.

By 1900, electric irons were available in the United States and Europe. A series of improvements to electric irons followed, including temperature settings suitable to various types of fabrics; a little water tank inside the iron that can be used to produce steam (which makes it easier to smooth out wrinkles); and a flat end so the iron can be set upright without the hot plate scorching the fabric covering of the ironing board.

100
B.C.

Edgar Degas' oil painting of women ironing

Cement

The Romans called it *opus caementicium*, from which we get the English word *cement*: the especially strong building material that can endure—often for centuries—the worst that nature has to throw at it. The recipe for Roman cement is 12 parts volcanic ash, 9 parts lime, 6 parts sand, and 16 parts gravel and small broken stones. Water is added and the mixture is stirred until it reaches a thick consistency, at which point it is ready for use at the construction site.

The most extraordinary cement structure built by the Romans is the dome of the Pantheon in Rome. Constructed as a temple to all the gods in 125 A.D., the dome is composed entirely of cement—5,000 tons of it. To accomplish this feat, the Romans used the cement in ingenious ways. First, as the roof of the dome rises to its apex, the thickness of the cement used diminishes: it is 21 feet thick at the base of the dome, but only 4 feet thick at the top. Next the Romans coffered the ceiling, meaning they used a pattern that resembles a series of sunken frames or boxes—this decorative motif also decreased the weight of the cement. The cement dome of the Pantheon has survived nearly 2,000 years of bad weather, barbarian invasions, warfare, and modern air pollution and it is still standing.

Modern cement differs from Roman cement. It is made by heating limestone and clay in a kiln until a rock hard lump known as a clinker is formed. The clinker is ground to a powder and mixed with a bit of gypsum to create what is known as Portland cement, the most common type of cement used today. Portland cement is also the primary ingredient in modern mortar and grout. Today, the world's top producer of cement is not Rome, but China. In 2006, China produced 1.235 billion metric tons of cement, which accounts for 44 percent of the world's cement market.

25 B.C.

The dome of the Pantheon is composed of 5,000 tons of cement

COMPASS

The Chinese called the compass a *sinan* and that version was a much more attractive device than the wobbly needle compasses that become popular in Europe. About 2,000 years ago, Chinese craftsmen discovered the magic of the lodestone, a magnetic chunk of iron ore. The magnetized stone lines up with the Earth's magnetic field, which runs north-to-south. The craftsmen carved the lodestone into the shape of a ladle. When placed on a smooth, polished stone slab or bronze plate, the ladle always pivoted with the handle pointing south and the bowl pointing north. Historians believe the Chinese chose the ladle motif purposely, as a tribute to the constellation known as the Big Dipper (or the Northern Dipper, as it is known in China), which is aligned with the North Star.

Sailors used the sinan to navigate; travelers on dry land also used a sinan to help them find their way. It's thought that China's jade miners, who wandered over thousands of miles of territory searching for fresh deposits of jade, always traveled with a sinan. Some scholars believe that the first sinans were originally made by highly skilled jade carvers.

By the eleventh century, simpler compasses were being used in China after scientists discovered that any iron needle would point north after it had been rubbed against a lodestone.

The compass had arrived in Europe by 1300 and may have been brought by Arab traders or come directly via European merchants, who for decades had been following the Silk Road to China. This was the "dry" compass—a magnetized needle enclosed in a small box and protected by a glass cover. There was also a "wet" compass, in which an iron needle was magnetized by rubbing it against silk and then floated on a small leaf in a bowl of water. The advantages of the dry compass were obvious: it was portable, all the working pieces were already in place, and there was no hunting around for a bowl, a leaf, an iron needle, and a piece of silk.

A fifteenth-century compass

EASEL

The prehistoric cave painters of France and the Roman fresco painters at Pompeii had no use for an easel because they painted directly on the wall surface. The easel did not develop, then, until a desire arose in the art world to paint pictures on panels that were portable (murals, of course, are not). In his writings on the history of art, Roman historian Pliny the Elder (23–79 A.D.) mentions seeing a large panel painting mounted on a wooden easel. A painted sarcophagus from about the same period shows an artist in his studio, and prominent in the scene is the classic three-legged easel artists still use today.

It appears that for several centuries the three-legged easel disappeared from the art scene, replaced by a high-sitting slanted desktop, usually attached to a high-backed wooden chair. Paintings in early medieval manuscripts show us artists painting sacred icons at these desks. While artists still painted frescoes on the walls of palaces and churches and public buildings, there was an enormous demand for small holy images painted on wood, which could be hung in the home, carried on one's travels, or even worn around one's neck. (These tended to be very small, comparable in size to the miniature portraits of the eighteenth century.)

One of the novelties of the Renaissance era was an interest in portraits. Fifteenth-century royalty, nobles, important churchmen and churchwomen, and even wealthy merchants commissioned the great masters to paint their portraits, often on a grand scale. The old medieval painting desk was much too small for such jobs, so the easel returned—also on a grand scale. To support these big paintings, artists built heavy-duty studio easels.

Small easels came onto the scene in the nineteenth century when artists preferred to paint outdoors. Lugging a ponderous studio easel to the top of a hill in the Hudson Valley or out to some field of sunflowers in Provence was out of the question, so artists' supply shops offered easels that were lightweight, portable, and even collapsible.

The final stage of easel development was the decorative easel, an elaborate piece of furniture upon which a family could display a prized painting or engraving. In the nineteenth century, some of the great furniture makers of the age, such as Charles Eastlake, designed splendid easels suitable for the parlor.

An early "desktop" easel

Vending Machine

Hero of Alexandria (10–70 A.D.), one of the cleverest inventors of the ancient world, devised a machine to dispense holy water at an Egyptian shrine: The pilgrim dropped a coin in the slot and a spritz of holy water came out of the tap.

Two thousand years later, in the 1880s in England, vending machines appeared that sold picture postcards. Richard Carlisle, a London publisher who also owned a bookshop, introduced a vending machine that sold books.

The first vending machines in America was manufactured in 1888 by the Thomas Adams Gum Company; placed on subway platforms in New York City, the machines sold Tutti-Frutti chewing gum. The classic gumball machine that dispensed round candy-coated chewing gum was invented in 1905.

Diners in Philadelphia got a treat in 1902 when the Horn & Hardart Automat opened its doors—it was the first completely coin-operated, vending machine, self-service restaurant in the world, and it stayed in business until 1962. In the 1920s, vending machines dispensed soft drinks in a paper cup. Today, vending machines sell everything from daily newspapers to sex toys. The country with the highest number and widest variety of vending machines is Japan. Japanese vending machines dispense bottles of liquor, women's underwear, fried food, even potted plants. It is estimated that there is one vending machine for every twenty-three people in Japan, yet the market is far from saturated. In 1999, Japan's vending machines generated $53.28 billion in sales.

50

Model Cindy Heller avails herself of a vending machine

DOME

The exterior of the Hagia Sophia in Turkey

Before the invention of the dome, there was very little open space inside large buildings. Instead, the interiors of such landmarks as the Parthenon in Athens and the Temple of Apollo at Delphi were a forest of columns: They blocked the view, but at least they held up the roof. Then, in about the year 100 A.D., Roman engineers were fooling around with a model of an arch. Someone turned it in a circle and suddenly the inspiration for an architectural element—the dome—was born.

One of the earliest domes ever constructed is still standing: 142 feet in diameter, the dome of the Pantheon covers the vast interior of the only intact Roman temple within the city of Rome. By the way, the Pantheon's dome was constructed using another Roman invention: cement.

An innovation in dome design came in about the year 532 in Constantinople. The Emperor Justinian was building an enormous new cathedral for his city, Hagia Sophia or Holy Wisdom, and he wanted a magnificent dome to crown the church. That was a tall order for the emperor's architects, but they discovered that if they carved 40 arched windows into the bottom, or corona, of the dome, it would lighten the load and enable them to build to the scale Justinian demanded. The windows also flooded the interior of the church with natural light, and created an unexpected optical illusion—the dome appears to be floating above the windows. The diameter of the Hagia Sophia dome is 110 feet—smaller than the Pantheon—but 210 feet high, which makes Hagia Sophia's dome 68 feet taller.

During the Italian Renaissance, tall domes became the fashion after Filippo Brunelleschi built a dome 375 feet high for the Duomo, or cathedral, of Florence in 1436. The Florence dome became the inspiration for the dome of St. Peter's Basilica in Rome, built by Michelangelo; it rises 394 feet. St. Peter's dome became the inspiration for the dome of St. Paul's Cathedral in London, designed by Christopher Wren in 1708. In turn, St. Paul's inspired the design of the dome of the U.S. Capitol in Washington, D.C.

The dome of the Hagia Sophia in Istanbul

Fork

The first eating utensil was not the knife or the spoon, it was the fingers. The ancient Egyptians, the Greeks, and the Romans all preferred to use their fingers, even after spoons and knives became available. The Romans even had a kitchen slave who cut their meat for them into bite-size pieces—this slave was known as "the scissor."

The fork appeared suddenly in the fourth century in the town of Byzantium, later Constantinople, known today as Istanbul. How it came to be is a mystery, but we do know that the Byzantines used forks in the early medieval era. In the eleventh century, when a Byzantine princess was married to the doge of Venice, she brought some little two-pronged gold forks with her. The Venetians were fascinated by the sight of their dogessa spearing her food and lifting it to her lips with this enchanting little tool. Some members of the Italian aristocracy aped the Venetian fashion and began to use forks at their meals—but by and large the utensil made little headway on the dining tables of Europe.

The first fork in England arrived with the writer Thomas Coryate, a souvenir of his travels on the Continent. Even so, well into the eighteenth century the people of the British Isles still scorned the fork. "Fingers were made before forks," sniffed the crotchety Irish clergyman and poet, Jonathan Swift, "and hands before knives."

None of the early English colonists brought a fork with them to America. The first record of a fork is 1630 when Governor John Winthrop of Massachusetts began to use one at mealtimes. The Americans were as reluctant to take up the newfangled utensil as their relatives across the pond.

The first forks had only two tines that were relatively widely spaced apart. As a result, small bits of food often slipped between the gap. In the late seventeenth century, the French added more tines so food would not drop through; they also made the tines slightly concave so the utensil could be used to scoop up food as well as spear it.

350

Early forks only had two prongs

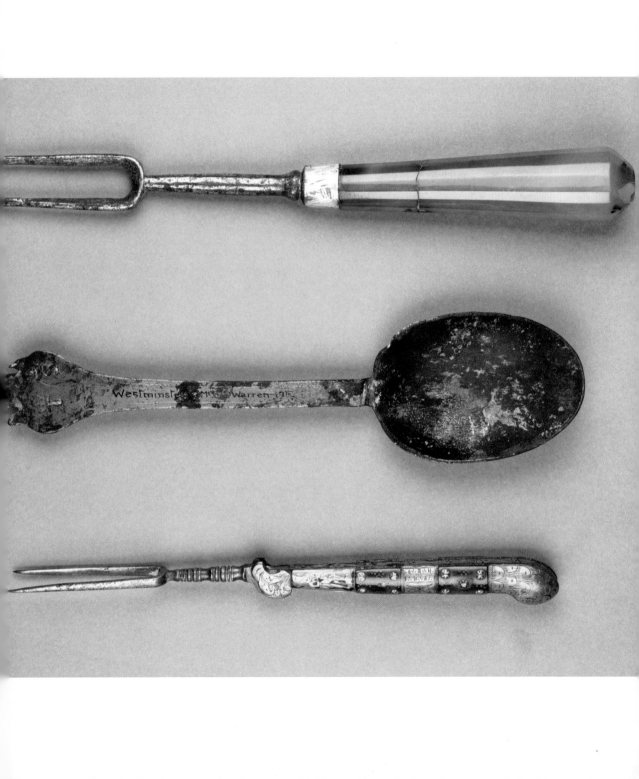

CAROUSEL

Also known as a merry-go-round, the carousel is a source for countless childhood memories. Its music, lights, and brightly painted animals and wagons have made it a favorite at amusement parks around the world.

The citizens of Constantinople were riding a carousel around the year 500, although they sat in baskets rather than on painted horses. The link between the horse and the carousel grew after the First Crusade (1096–1099). In the Middle East, Crusader knights discovered that Arab knights trained by doing battle with straw or wooden horses suspended from spokes at the top of tall pole. An animal or a slave turned the "carousel" so the knights could practice fighting a moving target. The Crusaders brought the concept back to Europe; in Spain, this training device was called a *garosello*, in Italy a *carosella*—both terms mean "little war," and so are the root of the name *carousel*.

The modern carousel of painted animals and elaborate wagons appeared at fairs and festivals in the early 1800s. One of the masters of these early carousels was Michael Dentzel, a German wagonmaker, who in 1837 gave up his trade and with his family became a full-time carver of carousel animals. The Dentzels also made their own mechanisms for their carousels so riders could enjoy the now-classic up-and-down motion.

The Dentzels sent some of their sons to America, where they opened a carousel-construction factory in Germantown, Pennsylvania, in 1860. The Dentzels were delighted to discover that Americans were crazy about carousels. To meet the demand, they brought over the best German woodcarvers and painters—but the market was huge and soon they had competition from carousel makers from as far away as Brooklyn and Kansas.

The late nineteenth and early twentieth centuries are considered the golden age of carousel construction when skilled European and American artists produced magnificent carousels. These carousel animals have made their way into art museums and are sought by private collectors.

November market at Hillerod, Denmark

Horseshoes

The practice of nailing metal shoes to a horse's hooves appears to have begun in Europe and may have been introduced by one of the barbarian nations that had overrun the Roman Empire, but the historical record is not certain. For example, Celtic graves of this period often contain the dead man's horse but no trace of the horse's shoes. On the other hand, there is a text from the Koran (written ca. 610) that describes warhorses "which strike fire, by dashing their hoofs against the stones." Iron shoes striking flint would give off sparks, of course, but the text may be metaphorical rather than a literal description of iron horseshoes.

We do know that about 500 years earlier, the Roman slipped "hipposandals" of a solid piece of metal attached with leather laces to their horses' hooves (*hippo* is the Greek word for "horse"), and riders in Asia had leather bootie-style shoes for their horses. By the Middle Ages, all horses wore iron horseshoes of the classic design of which we are all familiar nowadays. The purpose of the horseshoe is to protect the animal's feet and give it additional traction.

For centuries, all horseshoes were made by hand, one at a time. In 1835, Henry Burden (1791–1871) of Troy, New York, invented a machine that could produce sixty horseshoes in one hour.

Today, horseshoes may be made of titanium, aluminum, rubber, and even plastic.

In many societies, discarded horseshoes are thought to bring good luck, although opinions vary on how the shoe should be hung. Some believe the shoe should be hung over a door with the points downward, so the good luck will pour over you as enter and exit. Others insist that the horseshoe should be hung points up so the good luck doesn't drain out.

WINDMILL

In the first century A.D., the ingenious Hero of Alexandria invented a musical organ powered by a small windmill. It was a delightful toy. The first practical windmill may date from the reign of the Caliph Umar (ca. 584–644), who ruled in Medina and was a companion of the Prophet Mohammed. The large rectangular blades of this early windmill were turned by the wind and in turn moved a millstone to grind grain.

In the twelfth century, windmills sprouted up all over northern Europe. Previously, gristmills had been built beside streams or rivers, using the power of the current to turn the millstones. The windmills had the advantage of delivering more power than most watermills and they could be set up anywhere that caught a stiff breeze—including the fabled dry plains of La Mancha in Spain, where the author Miguel Cervantes had his hero, Don Quixote, tilt with the windmills.

By 1760, windmill technology had become wonderfully advanced. The blades could be adjusted to take the best advantage of anything ranging from a light breeze to an all-out gale. Inside the miller could adjust a regulator to keep the stone turning at a constant rate of speed.

The steam engine made the windmill obsolete but it got a second life in the nineteenth century on the arid and windy Great Plains of the United States: Ranchers erected windmills to pump water from deep underground wells. Today, entrepreneurs, environmentalists, and inventors are all trying to harness the wind as a cheap, inexhaustible way to generate electrical power.

Paper Money

The first currency was not paper but leather, specifically the hide of the rare white stag. China's coinage during the reign of Emperor Han Wu-ti (140–187 B.C.) had become unstable, thanks to private individuals who minted their own coins. With no uniform standards for coinage, China's monetary system was in chaos. To restore stability and confidence, the emperor issued the first treasury notes—a 1-foot-square piece of white stag hide stamped with a government seal; each piece of hide was worth 400,000 copper coins. It was a good idea in theory but in practice it proved to be difficult to execute because white stags were so rare and the more they were hunted down to be made into treasury notes, the more rare they became.

In 800 A.D., the Chinese issued the first paper currency. It was known as "flying money" because a good gust of wind could carry off a fortune. China's first paper money worked more like a check. A merchant who was about to go on a trading journey brought cash to a bank and was given in exchange a sheet of flying money which noted the amount the bank had received. When he reached his destination the merchant turned in the flying money to another bank for cash. The purpose of this transaction was to frustrate highway robbers. If a merchant was traveling with a certificate instead of cash, robbers would leave him alone.

Around the year 1000, the Chinese had a working currency system such as one finds in most countries today. Soon thereafter, Chinese counterfeiters got to work. One counterfeiter was caught after printing up 2,600 bogus pieces of currency; for his crime he was sentenced to death.

After their conquest of China, the Mongols adopted and disseminated many of China's innovations. When they rampaged through Persia in the late thirteenth century, they brought death and destruction with them, as well as the concept of paper money. It took longer for currency to take hold in Europe, where the first country to issue paper money was Sweden, in 1601.

Examples of paper currency issued by the Confederacy
during the American Civil War (1861–1865)

PLATE III.

CHIMNEY

At least since the Stone Age humans have huddled near a fire for warmth. The quest for fire may have been over but the issue of what to do about the smoke lingered for thousands of years. The simplest solution was to punch a hole in the roof directly above the fire. That is how the Vikings got smoke out of their wooden halls and how the American Indians got smoke out of their longhouses and wigwams. It was not an effective method, however, as the smoke usually drifted through the entire chamber as it floated up to the opening. Not to mention that during heavy rains and blizzards, it must have been difficult to keep the fire from being extinguished by the wet weather that poured in through the hole.

In 820, the Benedictine monks of the Abbey of St. Gall in Switzerland built fires beneath a clay pipe that drew the smoke up and out of the room. With this invention the chimney was born. The entire household would not have to gather around a single fire, choking amid the smoke; now it was possible for every room to have a fireplace with a chimney, thus making homes smoke-free and granting every member of the household privacy in warm, separate rooms when he or she desired it.

In most cases, chimneys have been built with brick, a sturdy, easy to use, fireproof material. On the American frontier where bricks were unavailable, pioneers lined the interior of their wooden chimneys with a thick coat of clay.

Among the most common improvements to chimney construction are chimney caps of perforated metal, which let the smoke out but keep birds, squirrels, and other small animals from coming down the chimney. Another useful innovation is the chimney cowl, a metal helmet-shaped object placed atop the chimney that changes direction with the wind and shields the chimney from gusts that would blow smoke and soot down into the house.

Gunpowder

It's often been the case in the history of science that while an inventor was trying to make one thing, he wound up stumbling upon something entirely different. A prime example is gunpowder: Sometime before the year 850 A.D., a band of Chinese alchemists were trying to find the recipe for an elixir that would grant long life, perhaps even immortality. They mixed saltpeter, sulfur, and carbon with honey, and they must have tried to blend the ingredients by heating them over a flame because an alchemy manuscript from the period warns its readers to not repeat the mistake of less cautious alchemists. "Smoke and flames result," the author wrote, "so that their hands and faces have been burnt, and even the whole house burnt down."

The mixture of saltpeter, sulfur, and carbon, of course, produces gunpowder (the honey was just a garnish).

By 919, gunpowder was China's secret weapon. It was packed into bamboo tubes tied to lances; the fuse was lit, the lance was hurled at the foe, and the explosion when the lance landed amid the enemy troops caused tremendous destruction, not to mention panic. A century later, the Chinese carried into battle both hand grenades—gunpowder packed into small pottery balls—and larger explosive shells that were thrown from catapults.

The Arabs and the Mongols acquired the secret of gunpowder, probably from Chinese prisoners of war. It was almost certainly through Arabic scientific books that an English Franciscan friar named Roger Bacon (ca. 1214–1294) first learned how to make gunpowder. Bacon was a philosopher and scientist who taught at the University of Paris. Once he started setting off gunpowder explosions, ignorant people whispered that Friar Bacon had acquired this wondrous and dangerous art by making a pact with the devil. The military men of Europe, however, realized the potential of gunpowder in battle and especially in siege warfare. Until this point, they had tried to batter down castles walls with stone missiles. Now, with gunpowder, they could blow the whole castle to smithereens.

American troops flee a grenade explosion

FIREWORKS

It stands to reason that China, where gunpowder was invented, would also be the birthplace of fireworks. The Chinese started with what we call a firecracker, but instead of a tiny paper tube and a miniscule amount of black powder, the Chinese packed gunpowder into a hollow chunk of bamboo and threw it into an open fire. The result is predictable—a burst of flame and a loud explosion. The Chinese considered these exploding tubes to be good for frightening away evil spirits, as well as celebrating festivals, marriages, or other happy events.

Around the year 1200, Chinese pyrotechnic engineers developed something they called "a ground rat." It was a hollow tube packed with gunpowder, as usual, but one end was left open. When the fuse was lit, the burning gas propelled the tube across the ground in an erratic manner. Civilians considered the ground rats entertaining to watch but the Chinese military saw another use for them—aimed at an enemy army, these burning, sizzling, skittering tubes would cause the infantry to break ranks and cause panic among the cavalry.

Marco Polo brought the formula for fireworks home to Venice in the thirteenth century and Italy became an early center of the thrilling new art. The Italian "fire masters," as they were called, introduced a number of improvements to Chinese fireworks. They attached the fireworks to rockets and shot them into the sky. They invented exploding shells that would burst high over the spectators' heads. They developed slow-burning explosives that gave off showers of colored sparks. And, building on the principle of the skittering ground rat, they invented fireworks that caused wheels to spin or fountains to spray.

Fireworks came to America with the early settlers in the seventeenth century. They were such a staple of public celebrations that, in 1777, to mark the first anniversary of the signing of the Declaration of Independence, the city of Philadelphia put on a display of fireworks.

Fireworks display to commemorate the peace of Aix-La-Chapelle. Handel composed his "Music for the Royal Fireworks" for the event in 1749

AMBULANCE

After the Crusaders took Jerusalem from the Saracens in 1099, they established a kingdom in the Holy Land. One group of knights opened a hospital in Jerusalem to care for pilgrims who fell ill during their visit to the holy places, as well as Crusaders who were wounded fighting the Saracens. Since they opened their hospital on the site of an old monastery dedicated to St. John the Baptist, the knights called themselves the Knights of St. John; but because they operated a hospital, they were better known simply as the Hospitallers.

To bring the sick or the wounded back to their hospital in relative comfort, the knights used a stretcher made of rope netting. The stretcher was attached to two long poles suspended between two horses—this was the first ambulance and it remained the standard for the next several hundred years. In 1793, Dominique Jean Larrey (1766–1842), Napoleon's personal physician, used horse-drawn wagons to bring the wounded from the battlefield to military hospitals. Like emergency medical crews today, Dr. Larrey's staff attempted to stabilize the condition of wounded soldiers on the battlefield before moving them behind the lines for more extensive treatment.

During a cholera epidemic in London in 1832, Dr. Larrey's methods were adopted for the civilian population. A horse-drawn ambulance picked up the sick at their homes and attendants began treatment during the ride to the hospital. But such ambulance services were temporary—when there were no battles or epidemics, the ambulance corps was disbanded.

The first professional, full-time ambulance corps was founded in London in 1887; taking the Hospitallers of the Crusades as their inspiration, they called themselves the St. John Ambulance Brigade. The first automobile ambulance went into service in Chicago in 1899, the gift of 500 city businesspeople who chipped in to purchase the car. During the Korean War, the United States Air Force used helicopters as ambulances—18,000 wounded were carried to hospitals in this way. Today, vans as well as helicopters are used as ambulances.

An early twentieth century ambulance

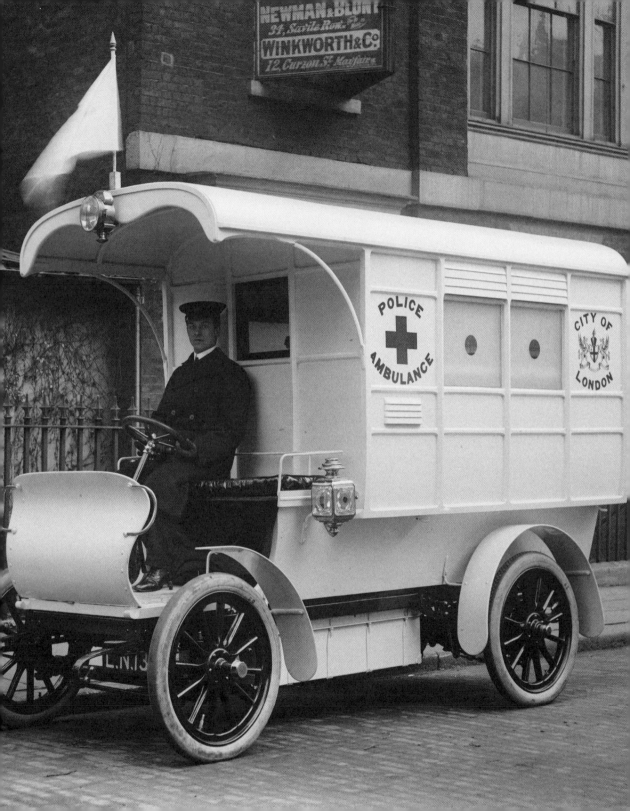

CANNON

The Tsar Cannon located at the Kremlin in Moscow, Russia

By the seventh century, the Byzantines had developed a primitive but effective flame-thrower to spew fire on their enemies. A small hand pump attached to the metal tube discharged the flammable liquid. It wasn't a cannon, of course, but the concept was in the ballpark.

Just as the Chinese were the first to discover gunpowder, they were also the first to make the most of it in battle. By 1128, they had invented the cannon. Archaeologists haven't found a twelfth-century cannon anywhere in China but they have found a remarkable sculpture in a cave temple in Ta-tsu, China, that shows a demon firing a small cannon; the sculptor even depicted the flaming projectile blasting out of the cannon's mouth.

Since the cannon would have to withstand repeated explosions, it was made of heavy metal—usually bronze. The earliest Chinese cannons were shaped like a modern light bulb—bulging at one end and then narrowing at the mouth. Gunpowder was packed in the bottom, a cannon ball of metal or stone was loaded in the piece, the fuse was lit, and the cannonball shot out.

An English manuscript dating from 1327 contains the earliest illustration in Europe of a cannon, which tells us that Europeans had started fashioning cannons before the book was illustrated. The manuscript illustration also shows the Europeans were using the same bulb-shaped model as the Chinese.

Over the next century and a half, kings and princes commissioned larger and larger cannons. One of the biggest was made for the Ottoman Turkish sultan, Mehmet the Conqueror. At the siege of Constantinople, he had a cannon with a mouth 35 inches across that could fire a 600-pound cannonball. On one occasion, the massive cannonball sailed past the walls of Constantinople and struck a ship in the city's harbor. The impact broke the ship in two.

Cannon being used at the siege of Constantinople.

Flying Buttress

On the night of June 10, 1194, lightning struck the great Romanesque cathedral of Our Lady in the town of Chartres, 50 miles from Paris. A tremendous conflagration burned for the next two days, destroying the church and most of the town. The townspeople were heartbroken, most particularly because they assumed that their most cherished relic, the tunic that the Virgin Mary was said to have worn at the Annunciation, had been destroyed in the fire. But, as work crews cleared away the debris, they found the reliquary with the tunic inside, safe and untouched by the flames. To the people of Chartres, the preservation of their relic was a miracle. To show their gratitude and devotion they vowed to build a new cathedral, more splendid than the last.

A new architectural style had just appeared in France characterized by high ceilings and tall pointed arches—the style we know as Gothic. The clergy of Chartres wanted their new cathedral to be in this new style but they insisted that the architect build on the foundations of the old Romanesque church. Measuring by the foundations, the cathedral was 427 feet long and 150 feet wide. The length wasn't a problem but the width would be—no Gothic church had ever been so broad.

So, an anonymous architect came up with an imaginative solution—the flying buttress. The thrust from the wide, vaulted ceiling would run down the exterior walls and through a series of strong yet decorative masonry supports along the entire length of cathedral. The architect discovered that the buttresses were so strong that he could cut out very large windows in the walls of the church and the structure would still stand.

From Chartres, the flying buttress spread across Europe, enabling architects from Dublin to Prague to build grand, soaring cathedrals filled with light from dozens of immense windows. Like the arch and the dome, the flying buttress is one of the world's architectural marvels, but we do not know the name of the man who invented it.

The Cathedral at Chartres, the first to utilize flying buttresses

HOURGLASS

The earliest surviving depiction of an hourglass is found in a painting called *Allegory of Good Government*, which Ambrogio Lorenzetti painted in 1338 for the city hall in Siena, Italy. In the painting, the hourglass represents the swift passage of time.

An hourglass is two glass bulbs, one above the other, connected by a narrow glass neck or channel and mounted in a wooden stand. The size of the bulbs and the amount of sand inside varies, depending upon how much time the hourglass is meant to calculate: There are 2-hour, 1-hour, half-hour, and even three-minute hourglasses (the three-minute variety is most often used for timing soft-boiled eggs). To "start" an hourglass you simply turn it upside down and wait for the all the sand in the top bulb to drain down into the bottom.

As a timekeeper, an hourglass is tedious—someone must attend to it constantly to turn it over when the sand has run out. Aboard ships in the fifteenth and sixteenth centuries, a boy was assigned this job. Most ships had 30-minute hourglasses; when the sand had run through, the boy turned the glass over and struck a bell to mark that another half hour had passed. Since no one wanted to keep a boy around the house or the shop simply to watch and turn an hourglass, it never became as popular as the sundial or the mechanical clock.

Instead of marking the hours of the day, the hourglass was used to delineate a precise period of time. For example, after the Reformation when the sermon became a major part of the religious service, some parish churches in England placed a one- or two-hour hourglass on the pulpit. When the preacher climbed into the pulpit he turned over the hourglass, then launched into his sermon; he was supposed to be finished by the time the sand ran out.

The one surviving use of the hourglass is as a kitchen egg timer. It takes three minutes for the sand to run out, at which point the cook has a perfect soft-boiled egg.

A sermon hourglass from St. Alban's in London

PRINTING PRESS

The closest thing to printing before Johann Gutenberg's (ca. 1395–1467/68) moveable type was carving a block of wood with an image or illustration, inking the inscribed surface, and pressing it against a sheet of paper. Early holy cards, or images, of the saints were produced in this way and were sold in shrine gift shops to pilgrims. Gutenberg's innovation was to cast individual letters on small bits of lead. These lead letters then would be arranged on narrow wooden trays or racks for each line of text. When the racks were locked inside a wooden frame, the result was a complete page of text. The lead was inked, paper or vellum was pressed onto it, and the result was a printed page. As many copies of the page as desired could be printed in this way, and when the job was done, it was simple to break down the frame and the trays and return the lead letters to their individual compartments until the next page was set up. Previously, all books, all written documents for that matter, had to be copied by hand. It was laborious, tedious, subject to frequent clerical errors, and expensive. Printing with moveable type, on the other hand, was simple, fast, and cheap.

In 1449, Gutenberg borrowed 800 guilders from Johann Furst (a considerable sum—enough to buy several farms) and opened his printing business in Mainz. To showcase his invention, he planned to print the entire Bible, a project which took years to complete. In 1456, his great printed Bible was done, but before Gutenberg could profit from his earth-shaking invention, Johann Furst sued to get his loan back—by now Gutenberg owed him 2000 guilders. Since Gutenberg did not have the cash, he was compelled to turn over his printing business. With Gutenberg's compositor to help him, Furst took advantage of the new technology and put out an exquisite Latin edition of the psalms. Furst made money from the first printing press but it is Gutenberg, the inventor, everyone remembers.

Gutenberg's printing press

GLOBE

One of the toughest historical myths to dislodge has been the assertion that Christopher Columbus was the only man among his contemporaries who believed the world was round. It simply isn't true. Only the most ignorant thought the Earth was flat; anyone with an education, especially scientists, knew the earth was round, going all the way back to the Greek geographer and astronomer, Ptolemy (ca. 83–161 A.D.).

While maps of the world were available throughout the Middle Ages, no one had ever thought to mount such a map on a sphere. The first person to do so was Martin Behaim (1459–1507) of Nuremberg. He was an astronomer and a cloth merchant. In the year 1480 or thereabouts, he traveled to Portugal where he worked at his cloth import-export business, while also studying the latest navigational and geographical discoveries. Throughout the 1400s, Portugal had been leading Europe in exploration of the African coast and the outlying islands in the Atlantic. After his marriage in 1486, Behaim moved with his wife to the Azores.

Early in 1492, Behaim returned home to Nuremberg for a visit. Under circumstances that have not come down to us, he decided to paste a map of the world on a round ball, thereby making the world's first globe. He called it an *erdapfel*—an earth apple. It is a perfect reflection of what well-educated people believed the world looked like in 1492, before Columbus sailed westward and ran into the Americas. There are some humorous errors—Japan is closer to the Canary Islands than to the coast of Asia, and the Atlantic Ocean is littered with mythical islands derived from ancient and medieval legends. Such quibbles aside, it is a splendid piece of work, and a rare historical treasure.

Behaim gave his globe to the city officials of Nuremberg, who put it on display in city hall. By the time Behaim died in 1507, the Europeans' picture of the world was much different and his globe was obsolete. Happily, Behaim's family, realizing what happened to obsolete maps and charts, took the globe back from the city and preserved it. It still survives and can be found Nuremberg's National Museum.

Screw

A factory specializing in the manufacture of tiny watch screws

The earliest screws were very large and carved by hand from heavy lengths of wood: They were used to crush grapes for wine, squeeze olives for oil, and even press wrinkles out of freshly laundered clothes. The earliest known metal screw with a notched head (which implies there must also have been a screwdriver by this time) was invented in Germany by an unknown clockmaker who used a file to cut the grooves, by hand, into a piece of wire.

In 1586, Jacques Besson (ca. 1540–1645), the inventor-in-residence at the court of King Charles IX of France, invented a machine to cut screws, thus bringing a degree of mechanical precision to screw-making that had not existed before. The output of Besson's machine was small, however. Mass production of screws came in 1760, when two English brothers, Job and William Wyatt, patented a screw-making machine that could churn out ten screws per minute.

A lathe for cutting screws was invented almost simultaneously in England and America: England's Henry Maudslay (1771–1831) brought out his machine in 1797 and America's David Wilkinson (1771–1852) introduced his in 1798.

The flathead screw was the standard until the beginning of the twentieth century. Automobile manufacturers found that flathead screws just didn't fasten the metal components of the car tightly enough. In the 1933, Henry Phillips (1890–1958) of Oregon designed what became known as the Phillips screw, with a cross-shaped indentation on the head. This screw could take more torque and so provided the tighter fit the auto manufacturers were looking for. Naturally, Phillips also invented the pointed Phillips screwdriver.

1513

Archimedes screw from the physics office of Jean-Jacques Rousseau

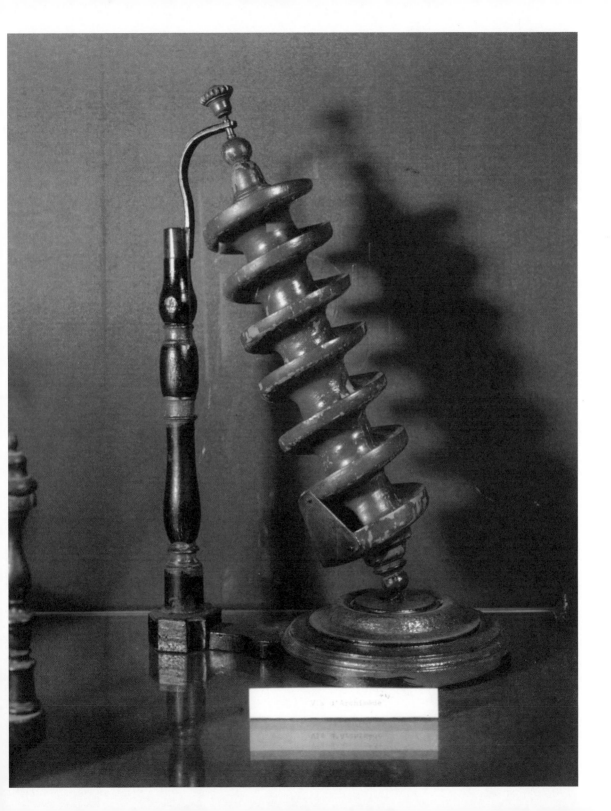

Vis d'Archimède

SAWMILL

Converting logs into boards was onerous work for most of human history. A typical method employed two strong men and a deep pit. After they felled a tree, they would cut off the limbs and branches and square the log. Then they locked it in place on a heavy frame built over a deep pit. The pitman was the one who climbed down into the hole; the topsawyer was, as the name suggests, the one who stood at the top, outside the pit. Using a whipsaw, a long, flexible saw with handles at either end, the sawyers used an up-and-down cutting motion to cut the log into boards, following chalk lines that had been marked out on the log before the sawing began. The stronger of the two was always the topsawyer since it was harder to pull the saw back up than it was to pull it down (the pitman may have been working in a pit but at least, when it came to sawing, he had gravity on his side).

The first sawmill that mechanized the work of sawyers was invented in 1593 by Cornelis Corneliszoon (1550–1607) of The Netherlands. Corneliszoon devised a simple arm and piston, attached it to a piece of technology the Dutch had in abundance—the windmill. The wind-driven machine Corneliszoon created used an up-and-down motion that imitated what the topsawyer and pitman did. Elsewhere in Europe, and eventually in the Americas, sawmills were built along swift flowing streams. In this case, waterwheels powered the saw rather than wind. The saw was the only thing mechanized in early sawmills, however, as men who worked in the mills pushed and guided the logs by hand.

Among the first places to build sawmills were the Baltic countries, the French colonies in Canada, and the English colony of Virginia. All three locations had immense forests—a natural setting for a lumber industry. By contemporary standards, the productivity of these early mills was low—only 500 boards a day—but by their own standards, after working in a pit with a whipsaw, 500 boards per day must have seemed incredible.

A mid–twentieth century sawmill

Newspaper

The *Times*. The *Post-Dispatch*. The *Inquirer*. Today's news hounds like to condense the names of their favorite newspaper to something short and punchy. But what did readers do with the world's first newspaper which carried the moniker, *Relation aller Fürnemmen und Gedenckwürdigen Historien* (*Collection of All Distinguished and Commemorable News*)?

The publisher was Johann Carolus (1575–1634) of Strasburg, an entrepreneur who recognized that news was a valuable commodity—and he was the man who could sell it. In an era when news was hard to come by, Carolus hired correspondents who were strategically located in the great cities of Europe. They sent him reports of the goings-on of the day, which Carolus compiled in handwritten newsletters. The newsletters were delivered to Carolus' subscribers, whom he charged a hefty sum for the privilege of being well informed.

It was a great way to make a living, except for writing out each copy of the newsletter by hand—that part was tedious and time consuming. In 1604, Carolus approached a printer's widow and bought everything in her husband's shop—from the lead type to the last sheet of paper. It was a masterstroke, the ultimate timesaving device that enabled him to increase his circulation and his income from subscriptions. In 1605, he released his first printed newsletter—what we would call a newspaper. But by October he had already run into trouble—other printers were getting their hands on his newspaper and reprinting it without his permission or paying him a royalty. He petitioned the Strasburg city council to intervene on his behalf.

The first newspaper in England, *The Weekly Newes*, began publication in 1622; the first American newspaper was John Campbell's *Boston News-Letter*, which first rolled off the presses in 1704.

Today, nearly one billion readers around the globe at least glance through a daily newspaper. Even so, the newspaper is in trouble: Most people get their news from the Internet, which can disseminate up-to-the-minute stories much faster than a physical newspaper.

Lloyd's List has the longest continuous history of any newspaper in the world

TELESCOPE

A reproduction of a rough sketch by Isaac Newton of an early telescope

In October 1608, representatives of the States General, the national government of The Netherlands, met to discuss two rival patent applications for "seeing faraway things as though nearby"—in other words, a telescope. The two men competing for the patent were Hans Lipperhey (died 1619), a German spectacle-maker who had settled in The Netherlands, and Jacob Metius (ca. 1571–1628), a Dutch lens grinder. Both men had mounted convex and concave lenses inside a tube that magnified far-off objects three or four times. Ultimately, the States General agreed that the invention was too easy to replicate for a patent to be granted to any one man, so they decided on a happy compromise: They awarded Metius a cash prize and hired Lipperhey to make several "spyglasses," as they called the telescope, for the government.

The Dutch council was correct—the telescope was easy to make and, in the next year, telescopes proliferated across Europe. In October or November 1609, Galileo Galilei made a telescope with a magnification capability of twenty times. Galileo became the first man to view the craters on the Moon, the four satellites of Jupiter, and the rings around Saturn.

Since Galileo's time, telescopes have become ever larger and almost all of them are pointed at the heavens. There is even a whimsical quality to the quest for the biggest telescope. The Paranal Observatory in Chile has four Very Large Telescopes (VLT), each of which have an aperture (the large opening at the end of the telescope) measuring 8.2 meters (26.9 feet). The European Southern Observatory (ESO) is planning an Extremely Large Telescope (ELT) with an aperture of 42 meters (137.7 feet). The ESO halted plans for the Overwhelmingly Large Telescope with an aperture of 100 meters (328 feet), due to the estimated billions of dollars it would cost to construct the monster.

A large telescope at the U.S. Naval Observatory

Barometer

A barometer predicts the weather by measuring atmospheric pressure. If the barometer reading indicates high air pressure, one can expect clear skies and fine weather; if the barometer reading indicates low air pressure, then a storm is brewing. By collecting barometer readings from a host of regions across the country, one can chart the movement of weather systems and make a reasonably accurate forecast.

The first mercury barometer was made in 1643 by an Italian physicist and mathematician, Evangelista Torricelli (1608–1647). He wanted to measure air pressure and initially had planned to use a tube of water for this purpose. When he ran the numbers, he realized that if he used water he would have to create a vacuum tube 60 feet long. So he switched to mercury, which is much more dense than water and required a vacuum tube only 3 feet long.

First, Torricelli drew out the air from his glass tube creating a vacuum. Then he sealed the tube at one end and placed the unsealed end of the tube in a dish of pure mercury. The weight or strength of the air pressure on the mercury in the dish forced it up the glass tube. He checked his barometer every day and noticed that the level of the mercury changed daily. Torricelli concluded that the mercury was responding to changes in the atmospheric pressure. Writing of the highs and lows of atmospheric pressure, he became a bit poetic. "We live submerged at the bottom of an ocean of elementary air," Torricelli wrote, "which is known by incontestable experiments to have weight."

Mercury, however, also responds to temperature. Consequently mercury barometers must be read in conjunction with a mercury thermometer. Altitude must also be taken into consideration when reading a barometer; the atmospheric pressure decreases at altitudes above sea level and increases at altitudes below sea level.

As of June 5, 2007, the sale of mercury within the member nations of the European Union has been restricted, so it appears that the almost 400-year history of mercury barometers in Europe has come to an end. As for mercury barometers in the United States, a few states have outlawed them, but there is no nationwide ban.

Navigational tools, including a barometer, from the early twentieth century

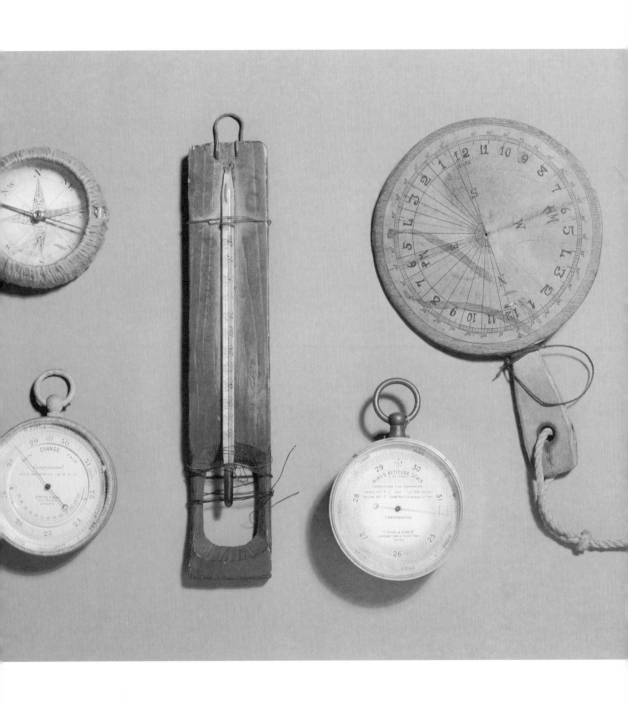

COMBINATION LOCK

The first locks date back at least 6,000 years, the earliest having been discovered in what is now Iran. For thousands of years, locks were opened or shut with keys. Such locks, of course, can be picked. It was not until the seventeenth century in England that an anonymous locksmith invented a new type of lock that was the prototype for the modern combination lock.

Inside the body of combination lock are a number of locking rings and at least three rotors or discs, as well as a device that enables the owner of the lock to reset the combination. The rotors are notched, with each notch corresponding to a number that appears on the numerical dial on the face of the lock. By turning the dial left and right to a series of numbers in the correct sequence, the notches on the rotors inside align perfectly and the lock opens. To secure the lock again, a person snaps the latch back into place and spins the dial, an action which misaligns the rings and rotors and makes it almost impossible for anyone who does not have the combination to reopen the lock. Because this type of lock is opened and closed by turning a dial, it is sometimes called a rotary lock.

Simpler forms of combination locks are found on chains to secure bicycles, as well as briefcases and luggage.

Although the combination lock cannot be picked, the combination can be cracked—hence the term "safecracking" (although the professionals prefer to call their art "lock manipulation"). Unlike what we see in the movies, it is a painstaking process. Yes, it requires a good ear (or a stethoscope) because each rotor makes a little clicking sound when it reaches its proper contact point. Since a very good combination lock will have eight numbers, the safecracker must make notes of each click before he can reproduce the correct combination, swing open the vault door, and cart off the loot inside.

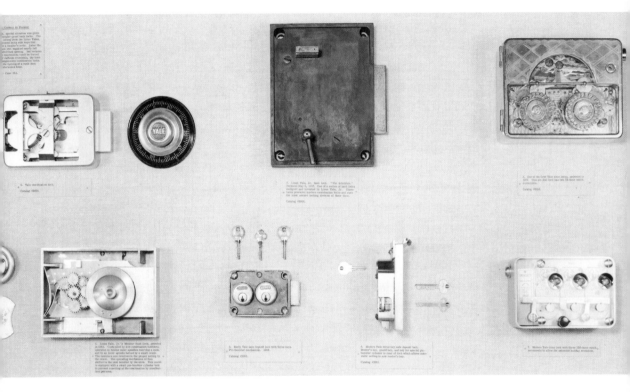

5. Yale combination lock.

Catalog #2654.

2. Linus Yale, Jr. Bank lock. "The dimension." Patented May 6, 1867. One of a series of bank locks designed and invented by Linus Yale, Jr. These varied prevented masters combination locks and overcome latest entirely locking devices of their time.

Catalog #2662.

3. One of the first Yale time locks, patented in 1877. This pin dial lock has two 24-hour movements.

Catalog #2663.

4. Linus Yale, Jr.'s Monitor Bank Lock, patented in 1861. Controlled by five combination tumblers, operated by fellow outer spindles carrying a cock, and by an inner spindle backed by a small crank. This operating mechanism controlled as it locked the proper setting by the crank. This operating mechanism of then started to the next movement by the crank. This model is equipped with a small pin-tumbler cylinder lock to prevent resetting of the combination by unauthorized persons.

Catalog #2651.

6. Early Yale safe deposit lock with Swiss keys. Pin-tumbler mechanism. 1869.

Catalog #2653.

4. Modern Yale drive-bar safe deposit lock. Member's key, guard key, and key for special pin-transfer cylinder in case of lock which allows automatic setting to new renter's key.

Catalog #2591.

7. Modern Yale time lock with three 120-hour movements, movements to allow for extended holiday weekends.

Microscope

During the first century A.D., as the Romans experimented with glass, making it in different shapes, some anonymous glassmaker made a piece of glass that was thick in the middle and thin around the edges. When he gazed through it, he noticed that this glass magnified objects. Another anonymous individual discovered that by catching sunlight through one of these glasses you could start a fire. These ancient magnifying glasses paved the way for microscopes 1,600 years later.

Early magnifying glasses enlarged the appearance of an object anywhere from three to ten times. They were essentially toys, often used to study small insects.

In about 1595, two Dutch eyeglass makers, Zaccharias Janssen and his father Hans, began experimenting with lenses. They mounted a few lenses in a tube and found that several lenses together would increase the size of an object well beyond anything a single magnifying glass could do. The Janssens had just invented the first compound microscope.

Antonie van Leeuwenhoek (1632–1723) was the son of a Dutch basket maker. When he was sixteen years old, van Leeuwenhoek was apprenticed to a Scottish cloth merchant in Amsterdam. The merchant had a magnifying glass and one of van Leeuwenhoek's jobs was to look through the glass to count the threads in a piece of woven cloth. In time, he saved enough money to open his own drapery shop.

In the 1660s, van Leeuwenhoek became interested in optics and began grinding his own lenses. He came up with a new design—small lenses, but with a pronounced curvature. When he looked through his lenses, he was astonished to find that magnification was 270 times an object's actual size! From this point on, van Leeuwenhoek moved away from the drapery business and began devoting more and more of his time to scientific pursuits. He mounted his lenses in a tube, making the first microscope, and began studying things that were, of course, microscopic. He was the first man to see bacteria, blood cells, and what he called "tiny animals" swimming in a drop of water.

Van Leeuwenhoek's microscope was a tremendous breakthrough for all of the sciences, especially medicine.

Early microscope

Steam Engine

The first steam engine designed and built in the United States, 1801

Thomas Newcomen's (1663–1729) calling was as a Baptist lay preacher, but he made his living selling tools and other metal equipment. Among his regular customers were owners of tin mines in Cornwall, who complained to Newcomen about the time and money they lost when the mines flooded. At the time, the only way to get water out of a mine was to put men to work using a manual pump, or hitching up a team of horses to haul bucket after bucket of water up from the mine. Both methods were time-consuming.

Newcomen invented a pump driven by a steam engine that would work around the clock and require neither a man nor a horse to keep it operating. Although the Greek mathematician Hero of Alexandria (10–70 A.D.) had discovered that power could be generated with steam, no one until Newcomen had actually found a practical use for a steam engine. In the nineteenth century, steam power would be harnessed to drive locomotives, steam ships, and factories; consequently Thomas Newcomen is often called the "Father of the Industrial Revolution."

Newcomen's method was actually simple: Water is heated in a sealed boiler and the heated water becomes steam. As the steam expands, it fills a narrow channel or cylinder connected to a piston. The pressure of the steam forces the piston to move, either in a pumping action or turning like a wheel. This pumping or turning drives the machine.

In 1788, John Fitch (1743–1798) operated the first steamboat on the Delaware River, carrying up to thirty passengers between Philadelphia and Burlington, New Jersey. Fitch's steamboat service went out of business because the same route was well served by good roads. In 1807, Robert Fulton (1765–1817) started a successful steamship company, making the run between New York City and Albany, New York.

A few years earlier Fulton had traveled to France in hopes of getting Napoleon interested in steam-powered ships. The emperor replied, "You would make a ship sail against the winds and currents by lighting a bonfire under her deck? Excuse me, I have no time to listen to such nonsense."

Advertisement for the first American steam engine railroad

The First Steam Railroad Passenger Train in America.

In 1826 a Charter was granted to the Mohawk & Hudson R. R. Co. for a Railroad to run from Albany to Schenectady, N. Y.; 16 miles. In 1830 work was commenced on the d, which went through the populous towns along the open streets, without restriction or fear of the consequences, and travelled across fields, up hill and down. The land was eithe n to the Railway Company, or sold for a trifling consideration, and it was finished in 1831. Both Locomotive Engines and Horses were used on the Road, and the tickets were sold tores or shops, or by the conductor, and the trains proceeded at a very slow rate. Stationary Engines were at the top of the hills, and the train was hauled up hill or let down hill by a ng rope. The brakemen used hand-levers to stop or check the train.

The first Steam Railroad Passenger Excursion Train in America was run on this Road in 1831. The Engine was named "John Bull;" it was imported from England—its weight wa ons. The Engineer was John Hampson, an Englishman. There were fifteen passengers on the Train of two coaches, among whom were the following,—(commencing at the rear o train.)

1. UNKNOWN.	6. MAJOR MEGGS, - - - Sheriff.	12. UNKNOWN.
2. LEWIS BENEDICT.	7. UNKNOWN.	13. EX GOV. JOS. C. YATES.
3. JAMES ALEXANDER. - - Pres't Commercial Bank.	8. BILLY WINNE - - - Penny Postman.	14. UNKNOWN.
4. CHARLES E. DUDLEY. - Dudley Observatory.	9. UNKNOWN.	15. UNKNOWN.
5. JACOB HAYES, - - - - - High Constable of New York.	10. UNKNOWN. -	16. JOHN HAMPSON, - - - Engineer.
	11. THURLOW WEED.	

THERMOMETER

For nearly 300 years thermometers, whether to register the temperature indoors or outside or the body temperature of a patient, have operated the same way: The mercury inside the thermometer expands when it comes in contact with heat (meaning the red column rises in the glass tube) or contracts when it comes in contact with cold (meaning it drops down in the glass tube). The first man to make a mercury thermometer was Daniel Gabriel Fahrenheit (1686–1736), a German physicist. The temperature scale he created to correspond with the movement of the mercury in his thermometer is known by his name—degrees Fahrenheit. At about the same time, a Swedish astronomer, Anders Celsius (1701–1744), devised another scale that used a 100-point scale to measure temperature, with 0 degrees Celsius being the freezing point of water and 100 degrees Celsius being the boiling point.

The once-ubiquitous glass thermometers filled with mercury are becoming harder to find. The dangers of the thermometer shattering and the potential for mercury poisoning have caused them to all but vanish from hospitals, doctor's offices, and home medicine cabinets. In their place is the digital thermometer, which is still inserted in the patient's mouth, and the ear canal thermometer, which was invented by a German surgeon, Theodore Hannes Benzinger, in 1964.

An ingenious gadget is the pop-up thermometer found in store-bought turkeys and chickens. The technology is wonderfully simple. Inside the plastic casing embedded in the poultry is the pop-up device with its point set in a piece of soft metal. This little rod or stick is wrapped in a tiny spring. The bird is ready to eat when it registers an interior temperature of 185 degrees Fahrenheit; at this temperature, the little blob of metal inside the thermometer's casing melts, which releases the rod and uncoils the spring, which causes the single-use rod to pop up.

Examples of thermometers made in seventeenth-century Florence, Italy

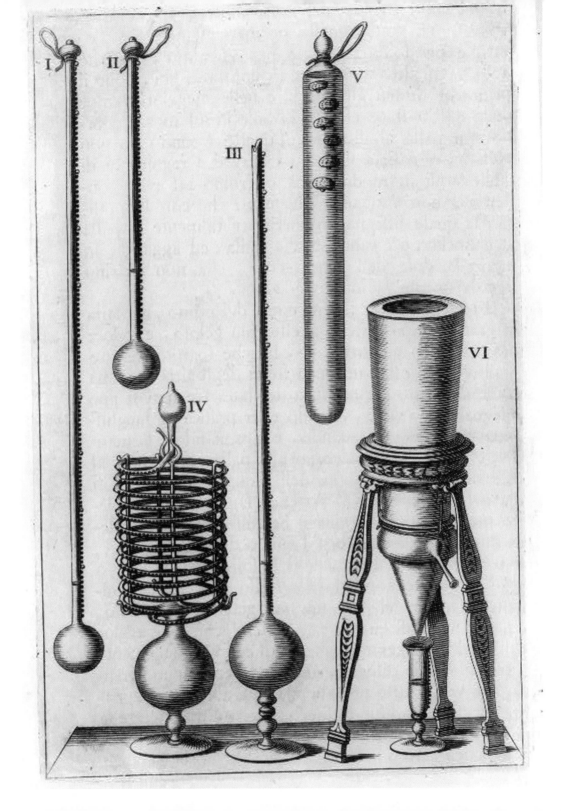

DENTAL BRACES

A pioneer in modern dentistry was the French dentist Pierre Fauchard (1678–1761). He devoted his medical career to curing gum disease, removing tooth decay, and creating false teeth. It was Dr. Fauchard who, in 1728, invented the first braces to correct the alignment of teeth. He also described his method in his book, *The Surgeon Dentist*, which included a chapter on how to straighten teeth. He used little flat strips of metal, which he tied to his patient's teeth with thread. A colleague named Bourdet published his own book in 1757, *The Dentist's Art*, in which he also recommended using metal devices to align the teeth. Fauchard and Bourdet are regarded as "the fathers of orthodontics."

The next stage of the development of braces came in 1841 when Dr. Schange used a screw to keep the braces in place. A major advance came in the late nineteenth century, when two dentists, Norman W. Kingsley and J. N. Farrar, wrote about irregularities in the teeth; Farrar recommended that the alignment of teeth could be corrected by using gentle force or pressure at regular intervals.

Metal brackets that could be adjusted to push the teeth into their proper place, in accordance with Farrar's principles, were invented in 1915 by Edward Angle (1855–1910) of Pennsylvania; his method remained almost unchanged until about 1970.

For many teenagers, wearing braces is a traumatic time. To diminish the self-consciousness or social stigma, transparent, removable braces, known as Invisalign, were invented by Zia Chishti and Kelsey Wirth during the 1990s; they went on the market in 2000. Completely transparent and not easy to detect, Invisalign appealed to many teenagers and found an unexpected market among adults who needed orthodontic work but were unwilling to wear metal braces.

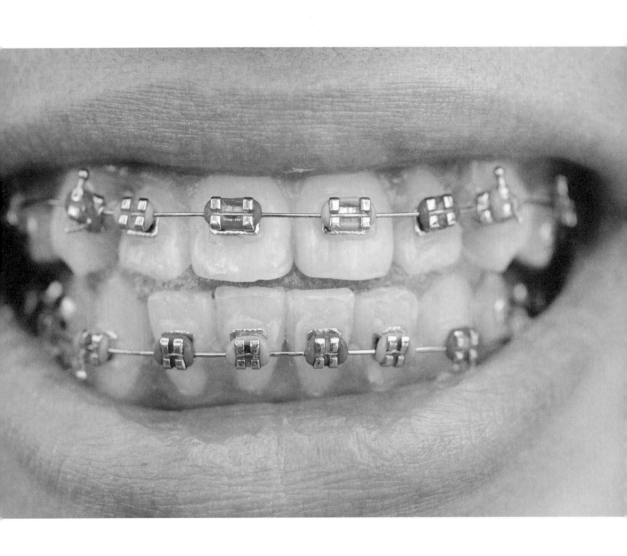

FLUSH TOILET

No, Thomas Crapper did not invent the flush toilet. True, he was in the plumbing business and his company did sell toilets but the mechanical flush toilet predates Crapper by more than a century.

As far back as prehistoric times, people used buckets of water to sluice away their waste. The earliest indoor toilet connected to a sewer system dates from 3000 B.C. and was built on Skara Brae in the Orkney Islands. In 1596, Sir John Harrington (1561–1612) designed and built a flush toilet for his cousin, England's Queen Elizabeth I, but the cascade of water made so much noise Elizabeth refused to use it.

The first valve-operated flush toilet (press down the handle and a valve releases a flow of water to flush the nastiness away) was invented in France by J. F. Blondel in 1738. This was followed by a steady stream of sanitary innovations and improvements. In 1775, a London watchmaker, Alexander Cummings, invented a trapdoor at the bottom of the toilet that kept a constant pool of water in the toilet, making it easier to carry away the waste and also keep out any foul odors from the sewer line. At England's Crystal Palace exhibition of 1851, public flush toilets were available for the first time. They were staffed by attendants who collected a penny from each user.

The toilet bowl is generally one of three varieties: a free-standing pedestal toilet, a toilet hung or mounted on a wall (a urinal), and the squat toilet bowl which is rarely found in the United States or Europe. While most toilets must be flushed manually, automatic flush toilets are found in public restrooms. Here there are two varieties: one flushes automatically on a pre-set schedule; the other is equipped with an electronic eye and flushes after the user has stepped away from the toilet.

1738

Nineteenth-century water closet

PATENT WASH-DOWN TRITON CLOSET.

MADE IN FINEST IVORY OR WHITEWARE PORCELAIN.

This ideal Closet possesses the advantages of the Valve Closet without its disadvantages.

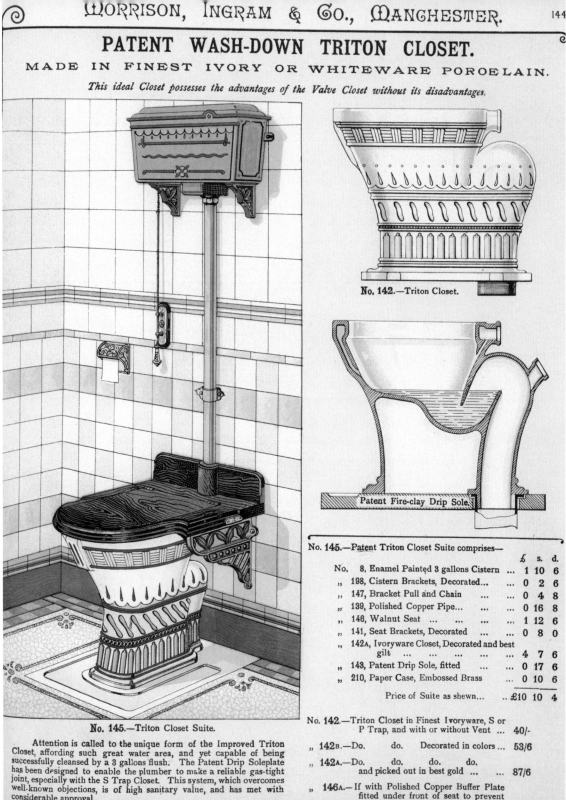

No. 142.—Triton Closet.

Patent Fire-clay Drip Sole.

No. 145.—Patent Triton Closet Suite comprises—

		£	s.	d.
No, 8, Enamel Painted 3 gallons Cistern ...	...	1	10	6
„ 198, Cistern Brackets, Decorated...	...	0	2	6
„ 147, Bracket Pull and Chain ...	...	0	4	8
„ 139, Polished Copper Pipe... ...	...	0	16	8
„ 146, Walnut Seat	...	1	12	6
„ 141, Seat Brackets, Decorated ...	...	0	8	0
„ 142A, Ivoryware Closet, Decorated and best gilt	...	4	7	6
„ 143, Patent Drip Sole, fitted ...	...	0	17	6
„ 210, Paper Case, Embossed Brass ...	...	0	10	6
Price of Suite as shewn...	..£10	10	4

No. 142.—Triton Closet in Finest Ivoryware, S or P Trap, and with or without Vent ... **40/-**

„ 142B.—Do. do. Decorated in colors ... **53/6**

„ 142A.—Do. do. do. do. and picked out in best gold **87/6**

„ 146A.—If with Polished Copper Buffer Plate fitted under front of seat to prevent urine trickling down front of Closet... **2/4 extra**

No. 145.—Triton Closet Suite.

Attention is called to the unique form of the Improved Triton Closet, affording such great water area, and yet capable of being successfully cleansed by a 3 gallons flush. The Patent Drip Soleplate has been designed to enable the plumber to make a reliable gas-tight joint, especially with the S Trap Closet. This system, which overcomes well-known objections, is of high sanitary value, and has met with considerable approval.

Franklin Stove

In eighteenth-century America, the only source of heat was a fireplace built into an exterior wall. With such an arrangement, the portion of the room closest to the fire was warm, while the rest was cold. Benjamin Franklin (1706–1790) of Philadelphia, arguably the most ingenious and most inventive American of his time, designed a cast-iron stove that could be situated in the middle of a room, radiating heat evenly into every corner. Furthermore, even after the fire went out or was extinguished when the household went to bed, the heat absorbed by the metal stove would continue to radiate heat.

However, there was a flaw in Franklin's design: He vented the smoke from the bottom of the stove, running a pipe up the chimney of a now-disused fireplace. This intensified the heat of the fire and consumed the fuel very quickly. Another Philadelphian, David Rittenhouse (1732–1796), a self-taught scientist and inventor, solved the problem. He added an L-shaped pipe to draw smoke up from the stove through an exhaust pipe that could be run up a chimney and out of the house. Thanks to Rittenhouse's innovation, the Franklin stove generated twice as much heat as a fireplace, but burned up only one-third as much wood—an especially attractive selling feature in eighteenth-century Philadelphia, at a time when the city's population was booming and firewood was expensive.

Franklin's stove became very popular yet he never filed a patent for it nor received any royalties for his invention. "As we enjoy great advantages from the inventions of others," he wrote, "we should be glad of an opportunity to serve others by any invention of ours; and this we should do freely and generously."

Diagram of Franklin's stove

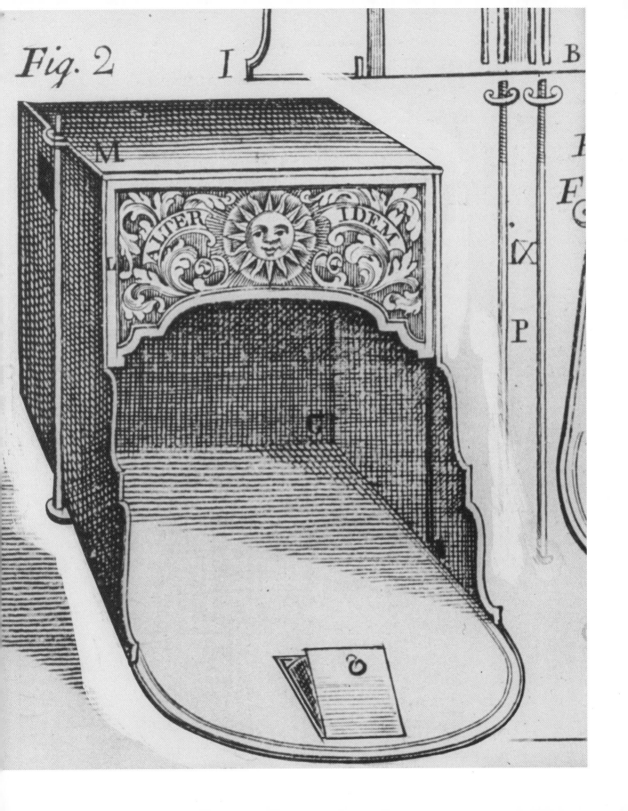

GLUE

Archaeologists have discovered 6,000-year-old broken clay pots that were pasted back together with tree sap. The first commercially produced glue of the modern era was manufactured in England in 1750 by reducing fish carcasses to a sticky mess. Since then, rendered animal bones, rubber, and starch have all been used to make glue.

Superglue was discovered in 1942 by Dr. Harry Coover of the Kodak Research Laboratories. He was trying to develop a clear, durable plastic for gunsights but by mistake made a substance called cyanoacrylate. It was unsuitable for gunsights but it had possibilities as an adhesive. Ultimately, Coover rejected it because he thought it was too sticky. Nine years later, Coover had relocated to Tennessee where he worked in the research department of the Eastman Company. With a colleague, Dr. Fred Joyner, he rediscovered the formula for cyanoacrylate. The scientists at Eastman recognized that Coover and Joyner's discovery had a practical application and the company began marketing the first superglue.

Perhaps the most popular glue is white glue, also known as school glue since just about every school kid has a bottle in his or her desk for classroom projects. This glue works best with porous materials such as paper, cloth, ceramics, or wood. It has many advantages—it's inexpensive, it dries clear, and it is not flammable. But, it is also not water resistant. Water can dissolve the bond of almost any object that has been pieced together with white glue.

Mid-nineteenth-century advertisement for Spaulding glue

D. SPAULDING & CO'S

UNRIVALLED
BUCKSKIN GLUE

LIGHTNING ROD

In the mid-eighteenth century, Benjamin Franklin came to believe that lightning is electricity. He based his conclusion on the color of the discharge, by the similarity between the erratic or jagged path of an electrical current produced in a laboratory and a bolt of lightning slashing across the sky, and the crackling sound both emit.

A common hazard of the time was lightning striking steeples, towers, and even ships' masts and burning the structure to the ground. If lightning was electricity, Franklin believed that a lightning strike could be neutralized. He developed a pointed iron rod 8 to 10 feet long and attached it to the tallest point of a building. One end of a metal wire was attached to the rod while the other end was buried in the ground. When lightning struck the lightning rod, the electrical current passed down the rod, into the wire, and fizzled out harmlessly in the dirt. Contrary to common misconceptions, a lightning rod does not ward off lightning, nor does it attract it; instead a lightning rod deactivates the destructive force of a lightning strike by detouring the electrical current.

If the metal used for the rod or the wire conducts electricity easily, there is a risk of massive heat damage to a structure. Franklin's iron rod and iron wire weren't bad but copper is a much better metal for conducting electricity. Modern lightning rods are usually made entirely of copper or at least have a copper point. The wire is now a copper cable, about 7/16ths of an inch in diameter. The cable runs down the side of the building directly into the ground—just like Franklin's metal wire did more than 250 years ago.

As was typical of Franklin, a man who looked upon scientific inventions as a form of public service, he never patented his lightning rod.

1752

A model of a courthouse, complete with lightning rod, attracts man-made lightning

BIFOCALS

It was the ingenious Benjamin Franklin (1706–1790) who invented the first pair of bifocals—for purely selfish motives. In 1764, Franklin was fifty-eight years old and having trouble reading. His condition was probably presbyopia, a loss in the eyes' ability to focus; it is a common occurrence as people age. Rather than go about with one pair of eyeglasses for seeing things at a distance and a second pair for reading or examining things close up, Franklin had two lenses ground and mounted, one on top of the other in a single frame. The lower lens, the one that was more convex, was for reading and other close work; the upper lens, which was less convex, was for seeing things that were more distant. Franklin may have been using bifocals before 1764 but that is the standard date given for its invention because, in that year, a cartoon was published that showed Franklin wearing his double-lens spectacles.

Since the lenses were separate, they were fragile and people who used bifocals often broke their eyeglasses. At the end of the nineteenth century, a French ophthalmologist, Louis de Wecker (1832–1906), improved bifocals by suggesting that the lenses should be fused together.

Over the years, the most common complaint heard from people who wear bifocals is getting accustomed to the line between the lenses that divides the field of vision. Many bifocal users complained of headaches and dizziness. To alleviate these complaints, a new form of bifocals known as *progressives* have come on the market. The new lens does not have a sharp line dividing the two fields; rather, the two lenses merge or blur together and the varying prescription strengths gradually change across the lens.

Benjamin Franklin wearing bifocal glasses, one of his many inventions

Jigsaw Puzzle

John Spilsbury of London was an engraver and mapmaker. One day, he had an inspired idea: He took a map of the world, glued it on a board, then, taking a fine-edged saw used to cut marquetry for furniture, cut out each country on the map. He marketed his new puzzle to schools as a painless method for teaching children world geography.

The jigsaw puzzle was soon recognized as a charming pastime, something that could be enjoyed in solitude or worked together with a group.

Throughout the nineteenth and into the twentieth century, jigsaw puzzles were wooden. Even today, wooden jigsaw puzzles are available for young children. Puzzles mounted on heavy cardboard became more common about 1900 after the invention of the die-cut process. This method uses thin, flexible, sharp strips of metal that can be formed into elaborate shapes, and mounted on a metal plate. The sheet of cardboard bearing the image passes beneath the plate, which is pressed down onto it. The sharp edges cut the cardboard, and the result is a jigsaw puzzle that can be mass produced. Such puzzles were economical—they sold for about 25 cents—which explains why manufacturers of wooden jigsaw puzzles didn't like them. The wooden puzzles sold for about $1.

Jigsaw puzzles tended to represent sentimental scenes, famous works of art, grand mountain scenery, or such thrilling new inventions as huge ocean liners or high-speed luxury trains. Later in the twentieth century, devilishly difficult jigsaw puzzles challenged the skills of puzzle masters—for example, a circular puzzle that was solid red.

1767

A 3D jigsaw puzzle of the globe

SUBMARINE

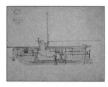

Design for an early submarine

David Bushnell (1740–1826) named his submarine *The Turtle*, although in shape it looked much more like a walnut. It stood 7.5 feet tall and 6 feet across and was built of oak and strengthened with iron bands. This was a one-man submarine, and the submariner lowered himself into the contraption through a hatch at the top then took his place inside on a stool. The submarine was powered by a propeller at the back; to move forward the operator turned a hand crank. As the submarine was heavy and there was also the resistance of the water to contend with, the man turning the crank had to possess good upper body strength.

Bushnell had designed his submarine to attack British ships (the American Revolution had begun in April 1775). He furnished the *Turtle* with a screw to penetrate the hull of a British ship, a mine that would be planted in the hole, and a fuse to set off the bomb. Once the mine had been set in place, the operator hand-cranked the submarine as fast as he could away from the doomed ship. Alas, the hulls of British ships were plated with copper, which Bushnell's screw could not penetrate; the *Turtle* never damaged, let alone sank, a single British vessel.

The first modern submarine was invented by John P. Holland (1841–1914), an Irish immigrant, who built his sub in Paterson, New Jersey. After experimenting with several prototypes, he achieved the greatest success with his Type 6 model: It held a crew of fifteen, was powered by a gasoline engine, cruised at a maximum speed of 45 miles per hour, and had a torpedo tube in its bow. Holland took it on its trial run in New York Harbor on St. Patrick's Day, March 17, 1898. Theodore Roosevelt, who was secretary of the Navy at the time, urged the United States government to purchase Holland's submarine but the government dragged its feet for two years before finally buying the Type 6 for $150,000.

Holland did not work exclusively for the Americans—he also sold submarines to the British and the Japanese. In 1905, the emperor of Japan presented Holland with the Order of the Rising Sun in recognition of the role his submarines played in Japan's naval victory over the Russian fleet in 1904.

Illustration of a Holland submarine torpedoing a ship

BALLOON

In eighteenth-century France, two brothers, Joseph-Michel Montgolfier (1740–1810) and Jacques-Etienne Montgolfier (1745–1799), became interested in gases that were lighter than air. Realizing that when air is heated it expands and becomes lighter than the surrounding colder air, the Montgolfier brothers decided to conduct an experiment—the first hot air balloon. On June 3, 1783, in the marketplace of Annonay, their hometown, the brothers brought out a linen bag that measured 35 feet in diameter. It was made of four separate pieces of cloth held together by 1,800 buttons and reinforced by a rope netting that slipped over the bag. They suspended the bag over a fire. The heated air inflated the bag, and when it reached its full capacity the brothers cut it loose. The balloon rose 1,500 feet in the air, then, propelled by wind currents, traveled 1.5 miles—in ten minutes.

On September 19, the Montgolfiers repeated their experiment in Paris, before a crowd of 300,000. Among the spectators was Benjamin Franklin. To make the demonstration more interesting for such a vast audience, the brothers attached a large wicker basket to the bottom of the balloon in which they placed a rooster, a duck, and a sheep they had named Montauciel (Climb-to-the-sky). This time, the balloon traveled 6 miles before the hot air gradually dissipated and the balloon came to gentle landing in the fields outside Paris. The animals were unharmed.

The first manned balloon flight occurred two months later when Jean Pilatre de Rozier (1756–1783), a teacher of chemistry and physics, and the Marquis D'Arlandes, flew 9,000 feet above Paris before landing 5.5 miles from their point of liftoff.

Inventor Jacques Charles (1746–1823) was the first to use hydrogen to inflate a balloon. Like the Montgolfiers, he conducted his experimental flights in Paris. Alas, when his invention landed in a peasant community, they attacked "the monster," completely destroying the balloon.

On a much smaller scale than these airborne versions, the first toy balloons were sheep or pig bladders, inflated with lung power the way most toy balloons are blown up today.

Illustration of the Montgolfier's first hot air balloon flight

GUILLOTINE

In most societies for most of human history, capital punishment was intended to inflict a painful, shameful death upon the victim. There were methods that would produce a swift, painless death for condemned members of the upper class but even these could go awry. Beheading with a sword or axe was fast if the headsman was skillful but there were exceptions. At the execution of Mary Queen of Scots, for example, the headsman was extremely nervous: his first blow missed the queen's neck and struck her head; the second blow sliced into the neck but not all the way through. It took three blows of the axe to completely decapitate the poor woman.

With the outbreak of the French Revolution, there was a consensus among the revolutionaries that a humane form of execution ought to be adopted. A member of the new National Assembly, Dr. Joseph Ignace Guillotin, recommended a new beheading machine he had heard of. "The mechanism falls like a thunderbolt," he said, "the head flies off, the blood spurts forth, the victim is no more." After building such a machine and testing it on cattle and corpses, the National Assembly adopted this method of decapitation as the only method of capital punishment permitted in France. The first victim was an armed robber, Nicolas Jacques Pelletier, who was guillotined before a curious crowd in 1792. In the two years that followed, the guillotine took the lives of more than 2,000 French men and women, most of whom were charged with the crime of opposing the revolution. The machine's most prominent victims were the king, Louis XVI, and his queen, Marie Antoinette.

Contrary to popular opinion, the guillotine was not invented by Dr. Guillotin; the first one was built by Dr. Antoine Louis of France's College of Surgeons. But, because of Dr. Guillotin's memorable speech in favor of such a machine, it was named after him and he has been credited mistakenly with inventing it.

The death of Maximilien Robespierre, one of the primary instigators of France's Reign of Terror

Battery

Fundamentally, a battery is just a little can of chemicals. The type of chemical used influences the lifespan of the battery. For example, zinc-carbon and alkaline batteries that are used in flashlights, remote controls, and smoke detectors will last for about one year. Lithium-iodide batteries last much longer, which makes them a good choice for such sensitive devices as pacemakers. And since lithium-ion batteries are rechargeable, they are used for cell phones.

There is one thing that all of these chemicals have in common: They can produce electrons for the electrochemical reaction that makes the battery work. Every battery has a negative terminal marked − and a positive terminal marked +. The electrons collect around the negative terminal. The battery starts to work once the electrons flow from the negative to the positive terminal.

The first electrical battery was invented in 1791 by Alessandro Volta (1745–1827), from whom we get the words *volt* and *voltage*. Volta discovered that he could create an electric current if he made a pile of alternating silver or copper disks (the positive terminal) and zinc disks (the negative terminal), and separating each pair of disks with a piece of cloth or blotting paper that had been soaked in salt water. While it was very exciting at the time, no one could figure out a practical use for the battery that would come to be known as "the Voltic pile."

The first practical application for batteries came about 1830 with the invention of the Daniell cell. It was a glass or clay jar. A copper plate was placed at the bottom, covered with a liquid solution of copper sulfate. Then a zinc plate was suspended in the jar and a liquid solution of zinc sulfate was poured over it. In the mid-nineteenth century, Daniell cells were used primarily to operate telegraph networks but they also found their way into private homes where they powered the first electric doorbells.

After dealing with piles of soggy plates and bowls of chemical solutions, the sealed battery was a blessing.

Depiction of various types of batteries, including the Voltaic

BATTERY.

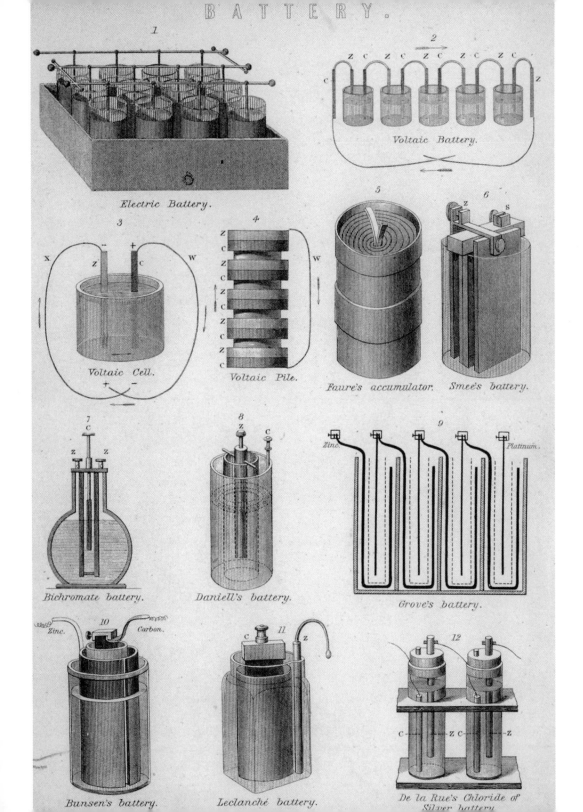

1

Electric Battery.

2

Voltaic Battery.

3

Voltaic Cell.

4

Voltaic Pile.

5

Faure's accumulator.

6

Smee's battery.

7

Bichromate battery.

8

Daniell's battery.

9

Grove's battery.

10

Bunsen's battery.

11

Leclanché battery.

12

De la Rue's Chloride of
Silver battery.

Cotton Gin

Newspaper illustration of a cotton gin from 1871

To answer the question everyone always asks—"gin" is a just a shortened version of "engine." Hence, the true name for Eli Whitney's (1765–1825) invention is the *cotton engine.*

Growing up in Westboro, Massachusetts, Whitney was a precocious boy adept with machinery. He learned the blacksmith trade and invented a machine to make nails. He wanted to go to college to study "the useful arts," as he called them but the colleges in America at the time did not teach such subjects. Eventually he took a degree in engineering from Yale, and later accepted a teaching job in South Carolina.

On the ship sailing south toward his new job, he met Katherine Greene, widow of the Revolutionary War general Nathanael Greene. They became friends and Mrs. Greene invited Whitney to visit her at her Georgia plantation, Mulberry Grove. So instead of going to the school, Whitney went to the plantation.

One of the crops grown at Mulberry Grove was green seed cotton. Although there was large market for cotton, it took the plantation's slaves so long to remove all the seeds from the cotton bolls that it was scarcely profitable. Whitney built a machine, an engine (a gin) that pressed the cotton against a screen while tiny metal hooks drew the cotton fibers through the mesh, leaving the large seeds to drop into a collection tray on the other side. There is an ongoing argument whether the idea for the metal hooks was Mrs. Greene's or Whitney's. More certain is that the invention made growing cotton, and having slaves to work the land, very profitable. A single cotton gin could produce 55 pounds of clean, seed-free cotton per day.

In 1794, Whitney received a patent for his invention but Mrs. Greene's neighbors broke into Whitney's workshop, studied the cotton gin, then went home to make exact replicas. Unable to defend his patent, Whitney left the South and moved to New Haven, Connecticut, where he gave up inventing and spent his life redesigning factories for mass production.

Eli Whitney's workshop, where he invented the cotton gin

CORKSCREW

By the 1600s, winemakers had discovered that by jamming a piece of cork into the neck of a bottle full of wine, the wine could be stored and allowed to age. The trick was getting the cork out. No one knows when the first corkscrew was invented. Historian Ron McLean of The Virtual Corkscrew Museum believes the first corkscrew was modeled on a tool known as a gun worm, a spiral length of metal used to clean gun barrels.

In 1795, an English clergyman, the Rev. Samuel Henshall (1764/1765–1807), received a patent for a corkscrew. It was a simple T-shaped corkscrew such as one still finds today. The handle was wood or bone. Descending from the handle was a length of metal, the top half of which was a smooth shaft while the bottom half was a screw. At the point where the shaft and the screw meet, Rev. Henshall introduced a small disk—known ever since as the Henshall button—to keep an overly enthusiastic wine lover from screwing straight through the cork. The cork was removed by getting a firm grip and yanking it out of the bottle, an exertion that requires between 25 and 100 pounds of force, depending on the condition of the cork (a moist cork tends to come out more easily, which is why a wine bottle should be stored lying on its side rather than standing upright).

An improvement on this design came in 1882 with the introduction of a corkscrew known as "the butler's friend." It resembles a pocketknife, with the screw folding against the handle. The innovation in this corkscrew is the small brace that rests against the lip of the wine bottle. When the screw is in the cork, you lift out the cork with the handle, using the brace for leverage.

Another common type of corkscrew today employs two levers. As a person turns the screw into the cork, the two arms of device rise up. By pushing the arms down the cork is forced up and out of the bottle.

The basic physics of the corkscrew have not changed since its invention

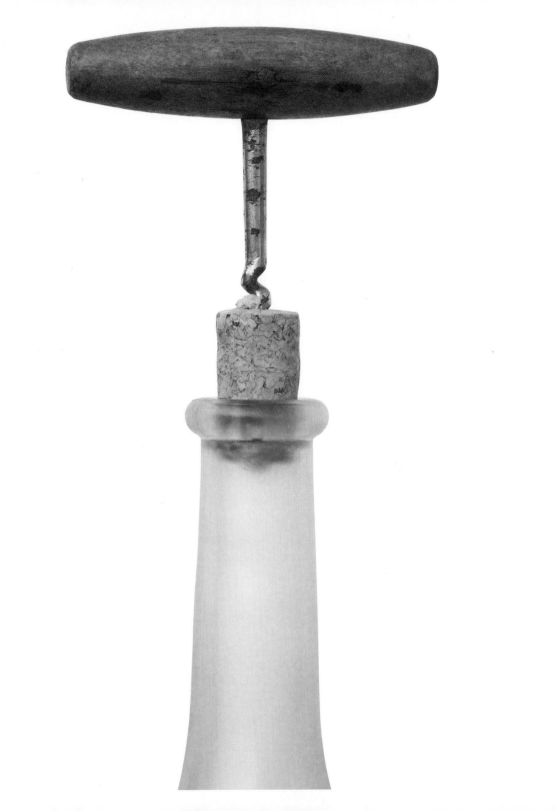

PENCIL

Graphite, a form of carbon, was discovered in the Seathwaite Valley near Keswick, England, around the year 1564. Soon thereafter, people in the neighborhood began using graphite in its raw state to write and draw. The pencil we know today was developed much later in France in 1795 by Nicolas Conte, a chemist. He mixed clay with graphite, fired it, and slipped it into a wooden cylinder. Conte discovered that the length of time the graphite and clay mixture in the kiln determined the hardness or softness of the lead (as we call it). Artists preferred a soft lead, while those who wrote wanted a hard lead that left clear marks on a page. One of the first American pencil makers was John Thoreau of Concord, Massachusetts; his son was the famous Henry David Thoreau.

In eighteenth-century England, if you wanted to erase an error you had made while writing with your piece of graphite, you rolled breadcrumbs over the page. By 1770, bits of India rubber were being used as erasers—but India rubber is a natural material that decomposes over time. It was Charles Goodyear who in 1839 developed the vulcanization process to make rubber last. In 1858, a Philadelphia man, Hyman Lipman, attached a vulcanized bit of rubber to the top of a pencil. His patent for his pencil-cum-eraser was later taken from him because the U.S. patent office felt he had not invented anything new, he'd just put two existing objects together.

Initially people used penknives to keep their pencils sharp. In 1847, a Frenchman, Thierry des Estwaux, invented the hand-crank pencil sharpener. Fifty years later, John Lee Love of Fall River, Massachusetts, invented the small, pocket pencil sharpener in which the pencil is rotated by hand. The Love Sharpener, as he called it, came with a little chamber that collected the pencil shavings for disposal later.

Cartoonist Thomas Nast sharpens a graphite pencil

HARPER'S WEEKLY.

A JOURNAL OF CIVILIZATION

Vol. XX.—No. 1040.]   NEW YORK, SATURDAY, DECEMBER 2, 1876.   [WITH A SUPPLEMENT. PRICE TEN CENTS.

Entered according to Act of Congress, in the Year 1876, by Harper & Brothers, in the Office of the Librarian of Congress, at Washington.

1876.
AMERICANS!

REMEMBER, WE ARE NOT UPON THE EVE OF A REVOLUTION.

REMEMBER, GENERAL GRANT *IS* PRESIDENT OF THE UNITED STATES.

REMEMBER THAT BUCHANAN *IS NOT* PRESIDENT OF THE UNITED STATES.

REMEMBER, THERE ARE NO TRAITORS IN THE CABINET NOW.

REMEMBER THAT IT IS THE COUNTRY WHICH IS AT STAKE, AND NOT GAMBLERS' POOLS.

REMEMBER THAT WE DON'T SCARE WORTH A CENT, AND IF HAYES IS ELECTED, HE *SHALL* BE INAUGURATED.

"The 'Solid South' has gone for TILDEN and HENDRICKS, and, by the God of battles, they shall be inaugurated!"—*Evansville (Ind.) Courier (Dem.)*.

"To see and dare and decide; to be a fixed pillar in the welter of uncertainty."

THOS. CARLYLE.

[That's U. S. GRANT.]

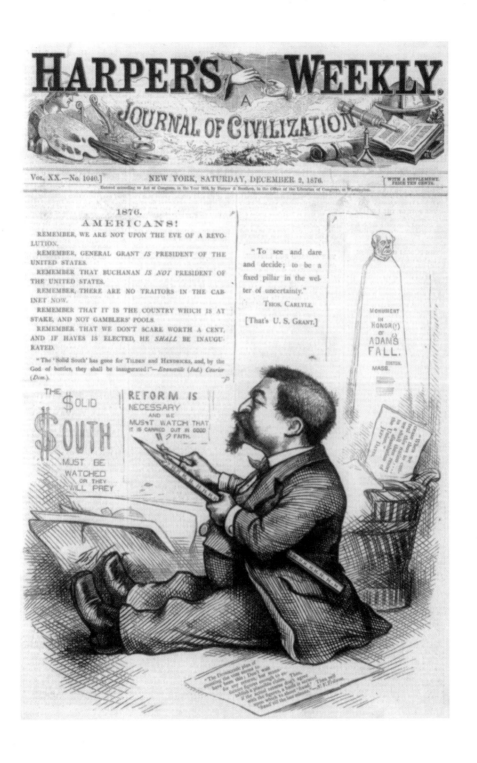

Vaccination

Vaccinations are vital for soldiers stationed overseas

One of the most dreaded diseases of the eighteenth century was smallpox. It is estimated that 60 percent of the world's population contracted the disease and about 20 percent died of it. Many of those who survived were badly disfigured by pockmarks—scars from the pustules that covered the smallpox patient's face.

Edward Jenner (1749–1823), an English country doctor, observed that dairy workers who contracted cowpox, a disease similar to smallpox but much less deadly, were ever after immune from smallpox. About the same time, a Dorset farmer, Benjamin Jesty (1736–1836) made the same observation. By taking pus from a cowpox blister, Jesty infected his wife and children. They developed a case of cowpox and never contracted smallpox.

In May 1796, a milkmaid named Sarah Nelmes contracted cowpox from a cow named Blossom. Jenner brought an eight-year-old boy, James Phipps (1788–1853) to Sarah, took some of the pus from her cowpox blisters and inserted it beneath the skin of both of James' arms. Days later after the symptoms of cowpox had subsided, Jenner inserted pus from smallpox blisters beneath James' skin—he did not contract the disease. Jenner's experiment on James Phipps is considered the first case of inoculation, the world's first vaccination.

Jenner's method of using a live virus as a vaccine is still employed. The flu shot many people receive before winter contains a live flu virus but a very weak one. Like cowpox, it immunizes the patient but only for one flu season (because flu viruses mutate constantly). When the patient's immune system finds this virus, it develops antibodies to kill it, then stockpiles these antibodies in case the virus returns again. It is these disease-specific antibodies that keep the patient immune from individual diseases.

Since Jenner's day, medical researchers such as Louis Pasteur and Jonas Salk have produced vaccines to prevent and even eradicate such diseases as rabies, polio, yellow fever, diphtheria, tetanus, measles, mumps, and, of course, smallpox.

A doctor administers vaccinations for smallpox

BALL BEARINGS

Anyone who has ever had to lug a heavy object knows that it moves much more smoothly, easily, and efficiently if it can be rolled rather than dragged. Dragging creates friction, which slows things down, but when something is rolled, friction is reduced significantly. That's the primary premise of ball bearings.

Typically ball bearings are set between two tubes or cylinders with some type of framing device (known as a cage) to keep the balls in place while preventing them from touching each other. Ball bearings are an essential component of countless motors and machines. If we did not have ball bearings, we would constantly have to replace component parts of machinery that had been worn out by friction.

Leonardo da Vinci is said to be the first to explain the uses of ball bearings, but the first patent for ball bearings was granted in 1797 to a Welsh carriage maker, Philip Vaughn, who used them in the axel assembly of his carriages.

Ball bearings come in different sizes to bear different types of loads. The ball bearings found in the wheels of a pair of inline skates or a lazy Susan are small. There are larger ball bearings for such things as conveyor belts and moving sidewalks in airports. Speaking of airports, giant ball bearings have been put to ingenious use in the San Francisco International Airport. Ball bearings 5 feet in diameter have been installed in concave bases beneath 267 columns that support the airport. If an earthquake strikes, the columns will roll over the ball bearings (each column is designed so it can roll 20 inches in any direction). Engineers hope that this ability to follow the motion of the earthquake will keep the airport's buildings from collapsing. After the earthquake is over, gravity will pull the ball bearings back to their original position in the center of the concave base.

Ball bearings were initially used on carriages

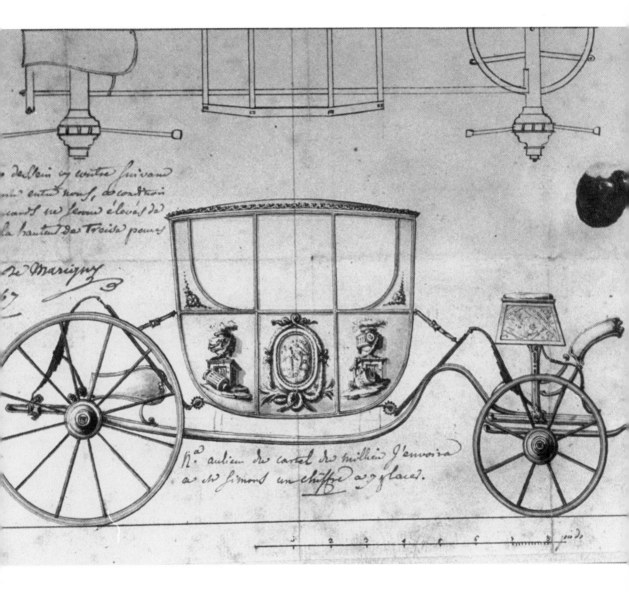

Parachute

Diagram of Garnerin's parachute

As was often the case with his ideas, Leonardo da Vinci imagined the possibility of a parachute and sketched some rudimentary designs for the thing but it would be other inventors who actually tried to make a working parachute. A Croatian inventor, Faust Vrancic, drew upon da Vinci's sketches to construct a rigid-frame parachute that he wore when he jumped from a tower in Venice in 1797. It worked, thank God, and Vrancic landed safely.

The first soft silk parachute was invented by André-Jacques Garnerin (1769–1823). He demonstrated it before an anxious crowd by leaping out of a hot air balloon as it drifted 3,000 feet above Paris' Parc Monceau. Garnerin landed safely and soon thereafter his wife, Jeanne-Geneviève, imitated his feat, becoming the first female skydiver.

The Garnerin parachute, when open, measured about 30 feet in diameter. It differed in one important respect from a modern parachute in that the jumper did not have the parachute strapped to his body, rather he stood in a little basket that was attached to the parachute. A parachute harness was not invented until 1887 by Captain Thomas Baldwin.

In 1890, two daredevils, Paul Letteman and Kathchen Paulus, perfected a method for packing a parachute in a knapsack strapped to the jumper's back, which was attached by a cord to the hot air balloon (Letteman and Paulus always jumped from hot air balloons). As the jumper leaped out of the balloon's basket, the attachment cord would snap and the neatly packed parachute would be released.

The question of who was the first to parachute from an airplane is under dispute—both Grant Morton and Captain Albert Berry claim to have made the leap in 1911.

Nylon replaced silk as the best material for parachutes

LOCOMOTIVE

The Japanese "bullet train" can reach speeds of 361 mph

The first locomotive debuted in 1804, courtesy of Richard Trevithick (1771–1833), a self-taught engineer who learned about machinery in the mines of Cornwall. He had discovered that if a steam engine released exhaust directly into the air, it would create a draft that drew a higher degree of heat from the fire thereby causing the boiler to generate more power.

Trevithick demonstrated his new invention in Wales, where there was already in place a track of iron rails: The iron miners in the region had laid these tracks, finding that it was easier for horses to pull wagons full of heavy cargo over rails than over bumpy, unpaved country roads. On the day Trevithick debuted his locomotive, it hauled 10 tons of iron ore and 70 men nearly 10 miles, from the mine at Penydarren to the village of Abercynon—this first train trip reached a maximum speed of 5 miles per hour. The trip took two hours. By 1808, Trevithick had improved his locomotive so it reached 12 miles per hour—he named it *Catch Me Who Can*. Sadly for Trevithick, his timing was off—his locomotive was too heavy for the cast-iron rails available in the first years of the nineteenth century and they cracked under the weight.

George Stephenson (1781–1848), an Englishman who also had spent his early years among miners, brought three great advantages to the advancement of the railways. With a chemist named William Losh he introduced wrought-iron rails, which were stronger than the brittle cast iron that had given Trevithick trouble. He set the width of a railroad track at 4 feet 8 inches (142 centimeters), which is still the standard gauge, and he mounted flanged wheels on his locomotive and railway cars so they would have a firmer grip on the rails. In 1825, he opened the first railway line for passengers—the Stockton and Darlington. The next year, John Stevens (1749–1838) introduced trains to the United States, laying out a circular track on his estate in Hoboken, New Jersey.

Stephenson's first successful locomotive

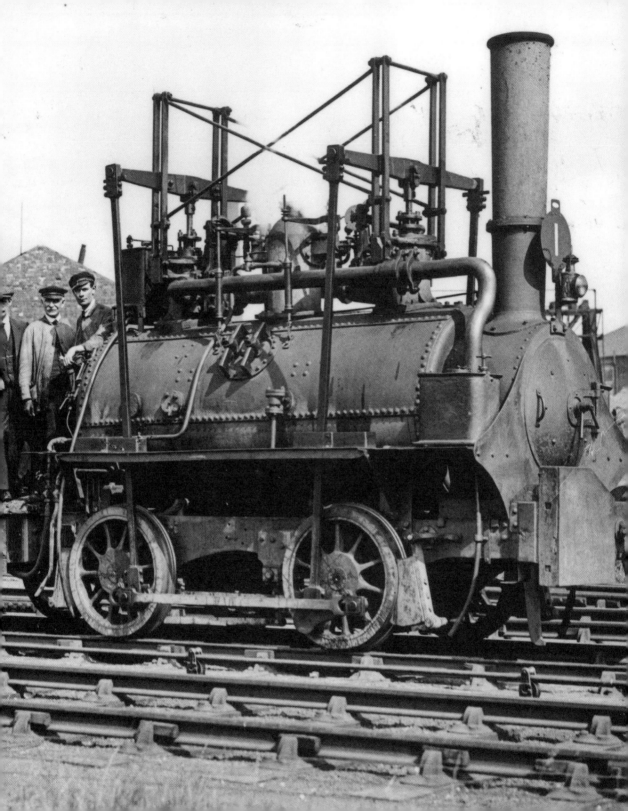

MATCHES

The first matches date back to the late Middle Ages and were lengths of cord that were lit and used to set off cannons and the first rifles. K. Chancel, an assistant in the Paris laboratory of the renowned chemist, Louis Jacques Thénard, invented the self-igniting match. Chancel coated the head of his match with a mixture of potassium chlorate, sulfur, sugar, and rubber. The match was ignited by dipping it into a bottle of sulfuric acid. Since it was both inconvenient and dangerous, Chancel's matches never found a large market.

An English scientist, John Walker (1781–1859), invented the friction match. He covered the tip of his wooden match with a mixture of antimony sulfide, potassium chlorate, gum, and starch, which would ignite when it was scratched over any rough surface. Walker called his matches "friction lights," but he never bothered to patent his invention. Walker's idea was taken up by Samuel Jones, who called his matches "lucifers": The word *lucifer* is Latin for "light bearer," and it is also a name for the devil. There was a hellish dimension to these early matches—they smelled terrible and they sometimes exploded upon being struck.

In 1836, a Hungarian chemistry student, Janos Irinyi, tried a new compound—phosphorus, lead, and gum Arabica. It worked beautifully; all the matches lit evenly without any explosions or foul odor. Irinyi sold his invention to a wealthy pharmacist, István Rómer, for about $120; Rómer made a fortune off the matches, while Irinyi died penniless.

All of these early matches were made of wood. The first stiff paper match in a paper matchbook was developed in 1892 by Joshua Pusey (1842–1906) of Pennsylvania. The designs on the exterior of antique and vintage matchbooks and matchboxes have attracted collectors; the hobby is known as phillumeny.

Tin Can

The Directory, the five-man executive committee that ruled France briefly between the end of the French Revolution and the rise of Napoleon (1795–1799), offered a prize of 12,000 francs to anyone who could discover an effective long-term method for preserving food. The winner was Nicholas Appert of Paris: He partially cooked the food, packed it into a glass jar sealed with a cork stopper, then plunged the jar into boiling water which expelled the air and airborne bacteria from the jar, thus keeping the food fresh. Appert sealed eighteen different types of food, including partridge, vegetables, and gravy, and sent them along with Napoleon's army who reported that everything had stayed fresh.

But glass breaks, so in 1810 an Englishman, Peter Durand, invented a tin can (actually it was an iron can plated with tin). Like Appert's glass jar, it was vacuum sealed and the food inside would stay fresh for months.

It was not Durand who opened the first tin can factory, however, but two other Englishmen, Bryan Donkin and John Hall. They began to mass produce food packed in tin cans in 1813. The British army and navy were Donkin and Hall's biggest clients. The navy alone purchased 24,000 large cans of food a year; especially welcome were the canned vegetables, which virtually eradicated scurvy among the ships' crews.

Yet another Englishman played a role in the history of the tin can: Thomas Kensett carried the tin canning process to America. He opened a factory on the New York waterfront where he canned oysters, lobster, salmon, and fruit. Explorers seeking the Northwest Passage and prospectors heading across the continent to the gold fields of California took tin cans of food along with them. During the Civil War, canned food was a staple among soldiers—it would have been almost impossible to supply fresh food to tens of thousands of troops in the field. It is estimated that between 1861 and 1865, soldiers ate approximately 120 million cans of food.

Tin cans were a vital part of a Civil War–era soldier's rations

STETHOSCOPE

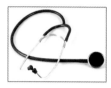

Rene Theophile Hyacinthe Laennec (1781–1826) was a French physician and a polite and proper gentleman. One day a young woman came to him with a physical complaint that, as part of the routine examination, would have required Dr. Laennec to place his ear directly against the woman's chest to listen to her heart. Embarrassed by the thought of such inexcusable intimacy, the doctor rolled up several sheets of paper into the shape of cone, placed the narrow tip in his ear and rested the mouth of the paper cone against the woman's chest. The sound of her heartbeat carried up the cone loud and clear. A year later, Dr. Laennec published a paper in a medical journal describing the success of this new method of physical examination, which he called "mediate auscultation"—a fancy term for "listening with a tool."

In his spare time Laennec tinkered at a little woodworking bench in his home. He fashioned the first device for mediate auscultation—a plain wooden tube with a funnel at one end and an earpiece at the other. He called it a "stethoscope," from the Greek words for "to examine the chest." Laennec was especially interested in the treatment of tuberculosis; he found that his stethoscope could be used to listen to beat of the heart, as well as the air passing in and out of the lungs.

Several physicians had worked up models of stethoscopes that permitted the physician to listen with both ears; Dr. George Cammann of New York created the first practical model in 1852. Cammann's stethoscope had ivory ear pierces attached to two metal tubes connected by a simple flexible hinge. The tubes connected to a hollow wooden ball (to amplify the sound), which sat atop a bell-shaped chest piece. It was elegant, simple, and useful—it enabled physicians to make more detailed diagnoses regarding the health of their patients. By the 1870s, the double stethoscope became the symbol of an up-to-date medical doctor.

A woman holding a model of Laennec's stethoscope

FIRE EXTINGUISHER

All fires need three things: high heat, oxygen, and fuel. In the most ordinary situation, when wood, paper, or cardboard are on fire, water will do the trick. But for burning chemicals, or burning electrical equipment that is plugged in to an electrical outlet, extinguishing the fire becomes more complicated. In those situations, you want something that will smother the fire, such as carbon dioxide which is heavier than the oxygen the fire needs to burn, or a dry chemical foam or powder that also will suffocate the fire. Most modern fire extinguishers contain either a dry chemical or carbon dioxide.

An Englishman, Captain George William Manby (1765–1854), devised the first modern fire extinguisher. It was a portable copper canister filled with 3 gallons of potassium carbonate solution; Manby's fire extinguisher used compressed air to spray the solution out of the nozzle.

An overhead fire sprinkler system was first installed in a New England textile factory in 1852. Suspended from the ceiling were a series of perforated metal pipes to discharge water throughout the room. There were two major disadvantages to this system: first, it had to activated by hand—someone had to turn the release valve that was connected the to the reservoir of water; and second, there was no way to limit the flow of the water only to the area that was on fire—once the water was flowing, the entire factory would get soaked.

Henry S. Parmalee, who owned a piano factory in Connecticut, invented an automatic sprinkler system. Parmalee designed a metal cap, which he soldered over the mouth of a water pipe. At 155 degrees Fahrenheit, the solder melted, the cap would pop off, and the water would spray the burning room. In 1874, Parmalee installed his automatic sprinkler in his own factory.

Frederick Grinnel of Providence, Rhode Island, purchased the rights to the invention from Parmalee and marketed it across the United States as "the fireman that never sleeps."

A motorbike fire engine from 1924 with fire extinguishers fitted to the sidecar

ELASTIC FABRIC

In 1820, Englishman Thomas Hancock (1786–1865) was inventing new types of gloves, shoes, and stockings that were fastened or held in place by using the stretching and clinging qualities of natural rubber. In cutting and shaping the rubber for these products, a great many small, useless bits were always left over. Hancock suspected that if the leftovers were ground up, melted over high heat, shaped and set out to cool, these bits of rubber could be useful again. To test his theory he built a simple hand-cranked machine—a wooden tube studded with sharp spikes that, when the crank was turned, minced up the odd pieces of rubber into smaller bits. Privately he called his machine "the masticator," which means "to chew up." Around other people he referred to it as "the pickle," so no one would have a clue what it did and could not steal his idea.

The following year, Hancock formed a partnership with Charles Macintosh (1766–1843), a Scotsman who had developed waterproof fabrics. They used the masticator to manufacture waterproof coats that became known as mackintoshes. Finally, in 1837, Hancock patented his invention (by now it was steam-powered and metal-built); it was no longer necessary to call it the pickle.

The next step in the process occurred in 1839 with Charles Goodyear's discovery of vulcanized rubber, a heated mixture of rubber, lead, and sulfur. While untreated rubber got sticky in hot weather and cracked in cold temperatures, vulcanized rubber was impervious to changes in the climate.

In 1876, Sir Henry Wickham, an Englishman, derived a new stretchy material from the sap of the rubber tree—latex. Like Thomas Hancock, Sir Henry used latex as a fastener for clothing. In the twentieth century, however, latex was made into form-fitting garments such as bathing suits, leotards, bodysuits, and surgical gloves.

Charles Macintosh, inventor of the Mackintosh raincoat

BRAILLE

The revolutionary system that enabled the blind and visually impaired to read and write evolved after Louis Braille (1809–1852) lost his eyesight as a result of a childhood accident. At age ten, Braille won a scholarship to the Royal Institution for Blind Youth in Paris, one of the few places where blind children could go for an education and to learn a trade. When Braille was twelve, a veteran of Napoleon's army, Charles Barbier, visited the school. During the Napoleonic wars, he had developed a writing system based on intricate configurations of raised dots and dashes for the troops to use as a secret code when passing sensitive intelligence to one another. Barbier's code flopped because it was too hard for most soldiers to learn but Barbier felt it might have some useful application, perhaps for the blind. Braille mastered Barbier's system but agreed that it was too complicated.

Three years later, at age fifteen, Braille introduced his own raised-dot reading and writing system. It was based on six dots that were arranged in different patterns to represent the different letters of the alphabet, as well as common punctuation marks. Braille's method worked so well and was so easy to learn that he went on to design a raised-dot system for musical notation (he was an accomplished organist) and for mathematical symbols.

Braille's fellow students at the school welcomed the new method for reading and writing but the teachers were hostile to it. First, they didn't like the system because the raised dots did not resemble letters, so the teachers couldn't "read" it just by looking at it. Furthermore, the teachers regarded Braille's method as a burdensome innovation that the school's administrators might compel them to learn. Opposition among the faculty was so strong that even after Braille was hired to teach at the Royal Institution, his method did not become part of the curriculum during his lifetime.

With Braille's death, it looked like his method would die out, too, but it was adopted in Great Britain and the United States, and from there it spread around the world. Later, Helen Keller wrote, "Braille has been a most precious aid to me in many ways. . . . I use [it] as a spider uses its web—to catch thoughts that flit across my mind."

Helen Keller with a Braille book

Lawnmower

Before there were mechanical lawn mowers, there were sheep. The gentry and the well-to-do kept private flocks of sheep to keep their vast lawns trimmed. In the late 1800s, the White House lawn in Washington, D.C., was still being manicured by hungry sheep.

But sheep were not dependable. They grazed in a haphazard manner, which meant parts of the lawn were neat and tidy, while in other parts the grass was tall and shaggy. In 1830, an Englishman, Edwin Beard Budding (1795–1846), invented the first mechanical lawn mower. He mounted blades on cylinder. As the machine was pushed forward, the cylinder rotated and the blades cut the grass. Some of Budding's earliest customers were the groundskeepers at the Regent's Park Zoological Gardens in London. In 1841, a Scotsman, Alexander Shanks, manufactured a 27-inch-wide lawnmower that was drawn by a pony rather than pushed by a human. The next year, Shanks came out with a 42-inch model that was drawn by a horse.

Human-powered lawnmowers were available in the United States in the nineteenth century but horse drawn lawn mowers were much more popular, largely because push lawnmowers were so heavy (they tended to be made of cast iron). Then, in 1870, Elwood McGuire of Indiana created a lightweight push-style lawnmower that proved to be a commercial success. Steam-powered lawn mowers appeared on the market in the 1890s and gasoline-powered mowers in 1902.

After World War II, with the exodus from the cities to the suburbs, gasoline-powered lawnmowers became standard equipment in every garage. The old-fashioned push lawnmower virtually disappeared from the market but they are making a comeback. Homeowners concerned about noise levels in the neighborhood as well as the use of fossil fuels are retiring their gas mowers and returning to the push variety.

1830

A Shanks lawnmower

MECHANICAL REAPER

Advertisement showing Cyrus McCormick in front of his reaper

At the time when Cyrus H. McCormick (1809–1884) invented his mechanical reaper, about 90 percent of the population of the United States was involved, to one degree or another, in agriculture. McCormick himself had grown up on a 532-acre farm outside of Lexington, Virginia, and he knew from personal experience the backbreaking labor involved in harvesting crops. One person sliced through the grain stalks with a handheld scythe, while behind him a second person tied the cut grain into bundles, which were tossed into a wagon and driven to the barn for storage. McCormick's father had tinkered with the idea of a mechanical reaper but nothing had come of it. Just before harvest time in 1831, Cyrus looked over his father's notes and began designing his own mechanical reaper. Within six weeks, he had built and tested his reaper. It was horse-drawn, so all the farmer had to do was drive back and forth across his fields while the mechanism held the stalks, sliced through them, then tied the stalks into bales.

To market his reaper, McCormick offered easy credit terms, guaranteed that with his reaper a farmer could harvest "15 acres a day," and made replacement parts for the reaper readily available. In 1847, to meet the demand for his reaper, McCormick opened a factory in Chicago.

He took his invention to Europe where he amazed audiences in England, France, Germany, and Austria. He won the Gold Medal at London's Crystal Palace Exposition for 1851, and he was elected to the French Academy of Sciences where his fellow inventors praised him "as having done more for agriculture than any other living man."

The mechanical reaper truly changed agriculture forever. Harvesting crops by hand had limited the size of family farms and, consequently, a farm's annual yield. With McCormick's reaper, harvesting grain was fast and simple. As a result, the size of farms increased and so did the amount of food available on the market.

Cyrus McCormick's mechanical reaper

Dry Ice

One day in 1835, Charles Thilorier, a French chemist, opened a canister of liquid carbon dioxide. Most of the carbon dioxide evaporated, leaving behind an extremely cold solid. He learned that when pure carbon dioxide changes from a liquid to a gas, the temperature in which the conversion takes place drops, causing some of the gas in the chamber to freeze. This solid residue is dry ice. Its temperature −109 degrees Fahrenheit—is much colder than standard ice made from water, consequently, dry ice will keep things colder than "wet" ice. And when dry ice is no longer needed, it can be left exposed to the atmosphere until it converts to gas again. Since it leaves no puddle or any other type of residue, dry ice is the tidiest, most environmentally friendly coolant on the planet.

Dry ice serves a wide range of functions. People use it to keep food fresh and for shipping medical supplies that must be kept refrigerated, such as blood for blood banks. The entertainment industry uses large blocks of dry ice to create fog in movies and stage plays. It can even be used to clean fuel tanks and large, industrial food processors.

Nonetheless, dry ice must be treated with care. Because it is so cold, if it comes in contact with bare skin it can cause severe frostbite and burn or tear the skin. Since it is carbon dioxide (the gas that we normally exhale into the air), as it changes from solid to gas, it displaces oxygen in the air. This can be dangerous if not fatal, therefore it should only be permitted to evaporate in a well-ventilated place or, better yet, outdoors. Finally, never place dry ice in an airtight container—when the dry ice converts to carbon dioxide gas the pressure could cause the container to explode.

A man holds a block of dry ice

COLT REVOLVER

Advertisement for Samuel Colt's revolver

In 1846, when the United States went to war with Mexico, representatives of the U.S. Army came to Samuel Colt (1814–1862) to place a large order for his repeating revolver. There was one problem: four years earlier Colt had been obliged to shut down his revolver factory, in large part because the army had taken no interest in the Colt handgun. What had changed the army's mind?

Before Colt's factory went out of business, the Texas Rangers had purchased a number of the revolvers for use in their war against the Native Americans. The advantage of being able to fire six rounds before having to reload was obvious to the Rangers, and they convinced the army brass that Colt's revolvers were just what the troops needed before they invaded Mexico.

Based on this recommendation from the Rangers the army placed an order for 1,000 revolvers; Colt promised that somehow he would deliver them. With no factory of his own, Colt approached Eli Whitney Jr., son of the inventor of the cotton gin. Whitney had a factory and he agreed to help Colt manufacture his revolvers.

Before Colt, rifles and pistols fired only one bullet at a time. After each shot, the gun had to be reloaded. (There were double-barrel pistols at the time, but firing two rounds was not exactly a great leap forward.) Colt's gun came with a revolving cylinder in which six bullets could be loaded—hence the name *revolver*.

The order from the army brought fresh demand for Colt's gun. In 1848, he opened a new factory in Hartford, Connecticut. In 1851 he went global, opening a factory in England in order to reach the European markets. Samuel Colt was soon one of the wealthiest arms manufacturers in the world. After Colt's death at age forty-seven in 1862, his widow, Elizabeth, took over the business, running it with great success until 1901, when she sold the company to investors. Elizabeth Colt died in 1905.

Portrait of Samuel Colt holding his famous revolver

Morse Code

An antique telegraph machine

Morse code, named for its inventor, Samuel B. Morse (1791–1872), is an international communication code comprised of "dots" and "dashes." The dots and dashes are a nickname for two types of signals used in transmitting messages via telegraph: a dot is an electrical signal of short duration, while a dash is an electrical signal of long duration.

Morse was a Massachusetts boy who became a portrait painter and a professor of art at New York University; he was also an amateur inventor and, like so many people of his day, was fascinated by electricity and looked for practical applications for this new source of power. Six years earlier another American inventor, Joseph Henry, had demonstrated that an electric current could be channeled through a wire to send a message. Henry set off the electric current at his end of the wire and it traveled more than 1 mile to its destination where it rang a bell.

Building upon Henry's research, Morse came up with the telegraph, which sent messages using his code of electric impulses. The electric current set in motion a small hammer which tapped out the dots and dashes on a long narrow strip of paper which a trained telegraph operator could read. Morse gave a public demonstration of his telegraph in 1838 but found no interested parties for his invention among the general public or the United States Congress.

Five years of persistent lobbying resulted at last in an appropriation of $30,000 from Congress to build a telegraph line 40 miles long, from Washington, D.C., to Baltimore. On May 24, 1844, the line was completed, and Morse and his supporters gathered in the old Supreme Court chamber in the U.S. Capitol to see him transmit the first telegraph message using Morse Code. He had asked the commissioner of patents' daughter, Anne Ellsworth, seven years old, to compose the first message. She chose a verse from the Bible, "What hath God wrought?" (Numbers 23: 23).

Samuel Morse operating a telegraph machine

BICYCLE

The bicycle, one of the most elegant modes of transportation ever invented, got off to a shaky start. The earliest model, built in 1839 by a Scottish blacksmith, Kirkpatrick MacMillan, had the pedals attached to the front wheel. The entire bike was made of wood, including the wheels. At the time, the bicycle was called the *velocipede*, which means "fast foot." More commonly it was referred to as the Boneshaker, a tribute to the bumpy ride the first bikers endured as they pedaled the wooden wheels over cobblestone streets.

Next came the large-wheel bicycle invented in 1864 by two Frenchmen, a blacksmith, Pierre Michaux (1813–1883), and a maker of baby carriages, Pierre Lallement (1842–1891). Once again, the pedals were attached to the front wheel, which was enormous but for a purpose—manufacturers discovered that the larger the wheel the more distance that could be covered by a single turn of the pedals. This bicycle (and for the first time the machine was known as a *bicycle*) had rubber tires that, along with the size of the front wheel and the long spokes, acted as a type of shock absorber, providing a much smoother ride. Undeniably striking and stylish, the large-wheel bicycle was also very expensive, costing the equivalent of six months' salary for an average working man or woman; consequently young people from well-to-do families were the primary market for the bicycle.

By the late 1880s, bicycles had been reduced to the size we know today. Biking had become such a popular activity that members of the League of American Wheelman (now the League of American Bicyclists) lobbied local government officials to pave their streets and roads for the sake of more comfortable cycling, especially bicycle racing.

In 1905, the course for Italy's Tour de Lombardy race passed a tiny chapel at the summit of steep hill called Ghisallo, overlooking Lake Como. The chapel was dedicated to the Virgin Mary and the racers began to invoke Our Lady of Ghisallo as their patron. In the 1940s, Father Ermelindo Vigano, an avid bicyclist, was assigned to the chapel. He petitioned the pope to make the patronage official—the request was granted in 1949 and Madonna del Ghisallo is the patron saint of bicyclists.

A man atop a Penny Farthing bicycle

CAMERA

So ubiquitous that it is now included in a cell phone, the camera is nonetheless one of the most astonishing inventions in human history. It does not just "take pictures," it chronicles the lives of individuals and families, enables people to see things that are thousands of miles away, and captures moments in time, from one moment ago or one hundred years ago.

Frenchman Louis Daguerre (1787–1851) is considered the true father of photography. In 1839, he revealed to the world his process for capturing an image: a copper plate was exposed to iodine fumes which made it sensitive to light; the plate was slid into a camera and the lens kept open for 10 to 20 minutes; the image was developed by warming the plate over heated mercury; the image was fixed permanently to the plate by immersing it in a warm saltwater solution; and finally, the plate was rinsed with hot distilled water. The astonishingly high quality of these "daguerreotypes," as Daguerre called them, still amazes us. There were a couple of disadvantages, however. Each image was unique—there was no way to make copies. And the surface of the daguerreotypes was so fragile that they were usually placed in protective cases under glass.

Cameras moved out of the studios of professional photographers and into the hands of amateurs in 1888 when George Eastman (1854–1932) of Rochester, New York, introduced an easy-to-use portable camera. There were no heavy metal or glass plates; the camera was loaded with a simple roll of film. All you had to do was point the camera at your subject, look through the viewfinder, and push the button. Once the film was used up, you dropped it off at any Kodak outlet (Kodak was the name Eastman dreamed up for his camera-and-film company) where the pictures could be developed.

The latest development in photography is the rise of digital cameras. These cameras contain a miniscule computer that records the gazillion colored dots (called pixels) that make up an image. While a traditional camera records an image on film, digital cameras record an image electronically.

A boy takes a picture using the first Kodak camera

POSTAGE STAMP

In the eighteenth and early nineteenth century, England had a peculiar postal system. The cost of postage for letters and parcels was determined by their weight and how far they were going, but it was the receiver, not the person who sent the mail, who was obliged to pay the postage. The British Postmaster General, Rowland Hill (1795–1879) suggested that postage rates ought to be uniform, they should be paid by the sender, and a stamp should be affixed to every letter and package to prove that the postage had been paid.

The English postal system adopted Hill's recommendations and issued the first postage stamp on May 6, 1840. It was a black-and-white portrait in profile of Queen Victoria, and the price of the stamp was one penny. It became known as the "Penny Black," and Victoria was so pleased with her portrait that she decreed all subsequent stamps printed during her reign (and she occupied the throne for sixty-three years) should use the image from the Penny Black. She also knighted Rowland Hill.

In 1843, Brazil and two Swiss cantons (provinces), Zurich and Geneva, adopted postage stamps. The canton of Basel followed suit in 1845. In 1847, the United States adopted postage stamps, printing two designs—a five-cent stamp with a likeness of Benjamin Franklin and a ten-cent stamp bearing a portrait of George Washington. In 1849, France, Belgium, and Bavaria issued their first postage stamps.

Sheets of perforated postage stamps were introduced by an Englishman, Henry Ascher.

Within months after the debut of the Penny Black, people were collecting it. In 1841, advertisements appeared in *The Times* of London asking readers to forward their cancelled stamps to ladies who wanted to the use the Penny Black to wallpaper rooms in their homes.

The "Penny Black," the world's first postage stamp

ANESTHESIA

The nineteenth century saw tremendous advances in medicine—including the use of clinical thermometers, the introduction of antiseptic methods, and the widespread implementation of vaccines. But there was one thing medical science had not overcome—pain. Patients who required surgery still faced a nightmare of pain. In an attempt to dull the agony of being fully conscious during surgery, physicians had tried everything from hypnotizing their patients to getting them drunk, but nothing worked.

In the late eighteenth and early nineteenth centuries, two English scientists, Joseph Priestly (1733–1804) and Humphrey Davy (1778–1829), suggested that nitrous oxide—better known as "laughing gas"—might anesthetize a surgical patient, but Priestly never brought his idea to the medical community; neither did Davy who, once he discovered the intoxicating quality of nitrous oxide, spent the rest of his brief scientific career inhaling the laughing gas and becoming thoroughly, hilariously stoned.

It was American physician Crawford Long (1815–1878) who studied the anesthetic possibilities of ether. He used ether successfully to anesthetize a patient in 1842 but he didn't get around to publishing his results until 1849 and, even then, only in a little obscure medical journal. While Long procrastinated, in 1846 another American physician, William Thomas Green Morton, used ether to put a patient to sleep before pulling the poor man's tooth. Dr. Morton, who had much better self-promotional skills than Dr. Long, had invited a panel of physicians from Massachusetts General Hospital to come and witness the procedure. Consequently, Morton, not Long, is usually credited as being the first to use an anesthetic successfully.

Only a year later ether had a rival. James Simpson (1811–1870), a Scottish obstetrician, used chloroform to alleviate the pain of childbirth. The new mother was delighted, but some prominent churchmen objected, citing Genesis as proof that God intended women to suffer pain when they gave birth. The debate was still raging when Queen Victoria, who was about to deliver her seventh child, instructed her physicians to give her chloroform to ease her labor pains. That settled the question, at least in the British Empire.

William Morton performing the first successful operation using ether as an anesthetic

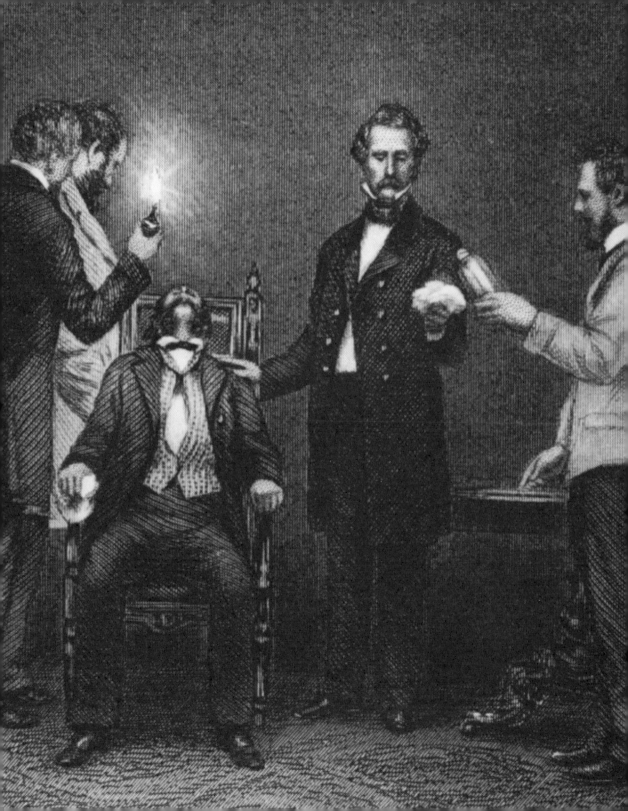

Vulcanized Rubber

Portrait of Charles Goodyear

Rubber in its natural state is a milky goo that oozes from certain tropical trees. There was a craze among entrepreneurs in the 1830s to find practical uses for rubber, and they used it in everything from waterproof clothing to life preservers. There was a serious problem with natural rubber, though: in cold weather it became rigid and cracked; in hot weather it melted into a stinking, sticky puddle; and in temperate weather it tended to stick to anything it touched. A Connecticut inventor, Charles Goodyear (1800–1860), realized that if rubber was to be useful it had to be stabilized.

In the 1830s, he began experimenting by mixing a variety of substances with rubber. His first breakthrough came in 1834 when he added nitric acid to rubber: It made the rubber firm, eliminated the sticky component, and gave it a smooth surface. Impressed by his improved rubber, the post office in Boston commissioned Goodyear to make rubber mailbags for their postal carriers. When summer came, however, the mailbags began to melt.

For the next ten years, rubber was Goodyear's obsession. He didn't hold down a steady job and, as a result, his wife and children were often on the verge of starvation. Several times he was imprisoned for debt. And he stank up the family home with his daily experiments in the kitchen.

His breakthrough came when he mixed sulfur with rubber and then heated it in a steam pressure cooker. The sulfur-and-steam-treated rubber remained firm; it was elastic but not sticky; and it was impervious to changes in the weather. In 1844 Goodyear received a patent for "vulcanized" rubber (he named his process after the Roman god of fire and volcanoes).

Goodyear's vulcanization process was so good that manufacturers in the United States and Europe stole his idea and began making vulcanized rubber without paying Goodyear any royalties. He spent most of the last sixteen years of his life in court defending his patent.

The Goodyear Tire & Rubber Company, by the way, was not founded by Charles Goodyear or any member of his family—it was simply named in his honor when the company began in 1898.

Charles Goodyear demonstrates his patented vulcanization process

Rubber Band

The year after Charles Goodyear discovered how to make a flexible, durable form of rubber, Stephen Perry, of Messers Perry and Co., patented the rubber band. This particular invention was another case of "the right place at the right time." Perry's company was among the first to utilize Goodyear's invention of vulcanized rubber by bringing rubber products on the market. Business was good but it generated a great deal of paperwork. As the documents piled up, the office looked increasingly untidy. Perry wanted some way to keep his papers together that was easier to use than tying them together with string or ribbon. In a moment of inspiration, he took a tube of rubber, sliced off a thin piece, and with one quick movement wrapped the circle of rubber around a pile of documents. The rubber band was simple, it made the pile of documents tidy, and it held them together more tightly that any ribbon or string.

The method of manufacturing rubber bands has not changed since Perry's day. A machine turns out a tube of vulcanized or synthetic rubber, which is sliced according to a preset width on a slicing machine. For larger or thicker rubber bands, the machine slices larger and thicker tubes of rubber.

While rubber bands are most often used to bundle objects together—everything from old bills to newspapers to fresh-cut flowers—children discovered years ago that rubber bands are also good projectiles. Some toy makers have introduced guns that fire rubber bands, but every school kid knows all you need is a thumb and index finger to shoot a rubber band across the classroom.

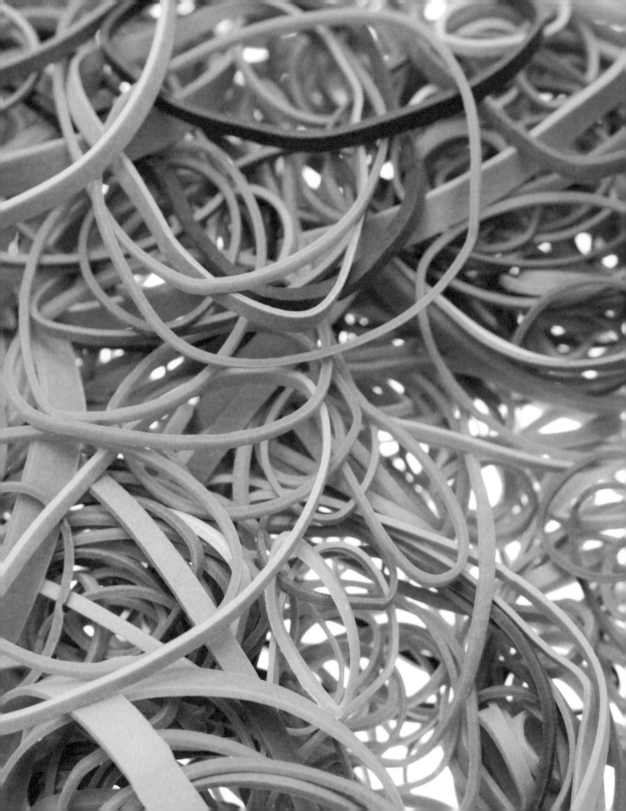

SEWING MACHINE

Elias Howe's sewing machine

The history of great inventions often brings with it the story of patent infringement. In 1845, Elias Howe (1819–1867) of Massachusetts put an end to thousands of years of painstaking drudgery by inventing the first practical sewing machine. Until Howe's machine came along, every piece of clothing, every shoe, slipper, or boot, every kind of fabric had to be sewn by hand. Howe's machine put an end to all that.

Unlike a traditional needle, the eye of the needle on Howe's machine was just above the point. By turning a hand crank, the sewer drove the needle through the cloth creating a loop of thread on the underside of the fabric where a shuttle slipped through a second thread and pulled the thread loop tight, forming a classic lock stitch. In a public demonstration, Howe's machine sewed more fabric than five seamstresses working simultaneously, yet no one bought his sewing machine. Armed with a patent issued to him in 1846, he went to England, hoping to market his invention there, but his design was pirated and he was swindled out of his royalties.

Howe came back to America where he found that Isaac Singer had also pirated his sewing machine. This time Howe went to court, suing Singer for patent infringement. The court ruled in his favor and in a deal negotiated between Howe and Singer and other sewing machine manufacturers, Howe received $5 for every sewing machine sold in the United States, and $1 for every sewing machine sold abroad. Once the sales figures had been totaled up, Howe received, in one lump sum, a check for $2 million.

Years of anxiety and financial hardship took its toll, however. Howe's wife died shortly after they returned to America, and Howe himself died in 1867 at only forty-eight years of age.

A nineteenth-century advertisement for Empire Sewing Machine Company

SUSPENSION BRIDGE

Photo of the completed Brooklyn Bridge

While studying engineering in Berlin, young John Roebling met the philosopher G. W. F. Hegel who told him America was "a land of hope for all who are wearied of the historic armory of old Europe." That sounded good to a young man eager to make a name and a fortune for himself in the field of engineering, and in 1831 Roebling boarded a ship for the United States.

While working his first job building canal equipment, Roebling invented a cable made of twisted wire to replace the hemp ropes used to haul barges up and down the canals. The wire cables were so strong and long-lasting that Roebling realized they could be used to carry enormous loads—a bridge, for example. Roebling imagined suspension bridges anchored by massive (but architecturally interesting) stone piers, one on either shore. The cables ran up to the top of the piers, then descended in a graceful arc between them. These cables bore the weight of the surface or deck of the bridge. Roebling's suspension bridges were incredibly strong, flexible in bad weather, and lovely to look at.

He built his first suspension bridge across the Monongahela River in Pennsylvania in 1846, then went on to build suspension bridges across the Kentucky River and Niagara Falls. His masterpiece from this period was his bridge across the Ohio River at Cincinnati—he constructed a bridge more than 1,000 feet long supported by a single swooping span of his remarkable wire cable. It is still in use today.

In 1869, Roebling received his most important commission—to link Manhattan and Brooklyn with a single-span suspension bridge. The ferryboat operators were livid at the thought of a bridge cutting into or even putting them out of business but the mayors of New York and Brooklyn were determined to have this bridge. Tragically, while supervising the early stages of construction Roebling's foot was crushed in an accident. The wound became infected and he died. His son Washington Roebling took over the job. However, in 1871, he spent a full twenty-four hours in one of the underwater foundation caissons of the bridge and suffered a near-fatal case of the bends that incapacitated him. From that point his wife, Emily Roebling, supervised the construction. The Brooklyn Bridge still stands—on completion in 1883, it was the largest suspension bridge in the world, and today it is a national historic landmark.

Opening celebration of the Brooklyn Bridge

Antiseptics

Two hundred years ago, the first thing a visitor to a hospital would have noticed was the smell—the stench of infected wounds that had become gangrenous. Usually these infections were caused by the hospital staff, who did not sterilize surgical instruments or even wash their hands before tending a patient's wounds. They were not callous rather, they simply had no idea that there was such a thing as germs.

It was an Austro-Hungarian physician named Ignaz Philipp Semmelweiss (1818–1865) who suspected that some perilous "material," as he called it, passed from doctors to patients. As the chief of obstetrics at the Vienna General Hospital, he ordered his medical staff to wash their hands with chlorinated lime, a solution that can also be used as a bleach, before examining a pregnant woman or assisting at the birth of a child. The medical community mocked him, but Semmelweiss compelled his staff to follow his instructions. As a result, the number of mothers at Semmelweiss's hospital who died of the infection known as puerperal or childbed fever dropped from 12.2 percent to 2.3 percent.

Meanwhile, at the Glasgow Royal Infirmary in Scotland, a young surgeon named Joseph Lister (1827–1912), inspired by what he had read about Dr. Semmelweiss's work, began experimenting with carbolic acid in the hope of reducing cases of gangrene after surgery. He found that wounds cleaned with carbolic acid were much less likely to become infected. Encouraged by this initial result, Lister had all his surgical instruments washed in carbolic acid, insisted that his surgical team wash their hands in carbolic before and after surgery, and even had carbolic acid sprayed all around the operating room. In 1867, the British medical journal, *The Lancet*, published Lister's finding in a paper entitled "Antiseptic Principle of the Practice of Surgery." It proved to be a groundbreaking work that revolutionized medical care around the world.

Dr. Lister became so closely associated with eliminating germs that a manufacturer of an antiseptic mouthwash named their product after him—Listerine.

1847

Carbolic antiseptic spray being used to prevent infection

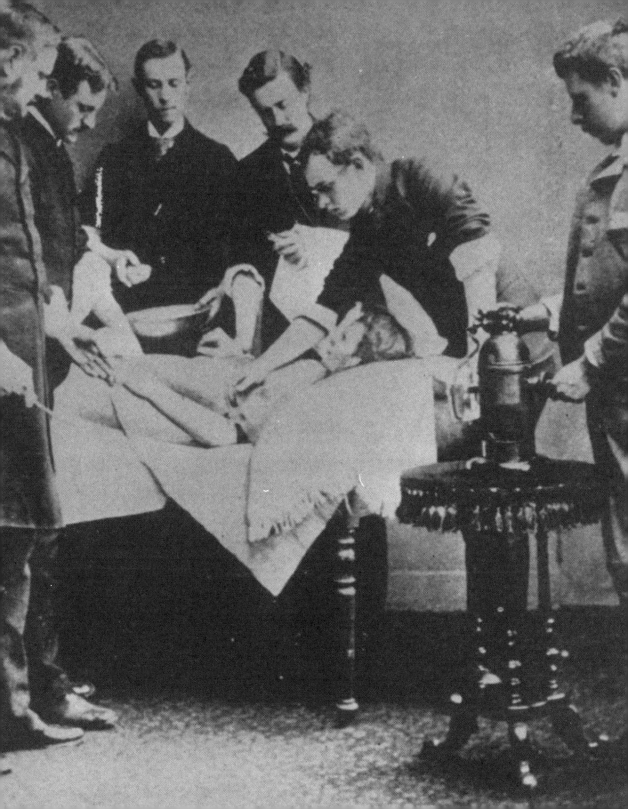

ODOMETER

William Clayton (1814–1879) was among the first Mormons to set out for Utah. The Mormons' leader, Brigham Young, had given Clayton responsibility for calculating how far their wagon train traveled each day. Clayton determined that a wagon wheel had to turn 360 times to cover one mile, so he tied a red rag to a wheel and counted each revolution. At the end of the day, he divided the number by 360 and knew how many miles the pioneers had covered. It worked but it was mind-numbingly dull and there was a very high chance of miscounting. He needed something more reliable.

Fortunately, among the settlers were a mathematician, Orson Pratt, and a skilled carpenter, Appleton Milo Harmon. The three men put their heads together and invented a device they could mount over a wagon wheel. A bar counted each revolution of the wheel. The bar was attached to a gear that moved one tick forward at each quarter-mile the wagon covered. At the end of the day, Clayton could determine how many miles they had traveled by observing how far the gear had advanced. To protect their invention from road dust and rain, Harmon built a wooden case that covered the device. They called it "the roadometer."

Clayton's mile-counter is still the basis for mechanical odometers. In an automobile, a mechanical odometer is hooked up to a tiny gear that spins as the car moves along the road. It spins 169 times to register one-tenth of a mile on the odometer, or 1690 times to register a mile.

It is said that the Roman engineer Vitruvius mounted a wheel 4 feet in diameter on a frame. He determined that it took 400 revolutions of his wheel to cover a mile. To the axel of his wagon, he connected a gear with 400 teeth and a pin. At each rotation of the wheel, the pin advanced the gear one notch. When it reached the 400th notch, a pebble dropped into a container. By counting the pebbles, Vitruvius would know far he traveled. It sounds like Clayton's roadometer but there is no evidence that Vitruvius ever built his device.

Mormon pioneers coming over Little Mountain, 1847

SAFETY PIN

Readers of a certain age will recall the days when diapers were cloth and safety pins were used to keep the front and back of the diapers together. The man who invented the safety pin, Walter B. Hunt, made it by accident. In the first months of 1849, Hunt was $15 in debt and had no way to pay back the money. He sat at his desk, running through his mind one scheme after another to raise enough cash to get out of his current financial hole and, as he worried, he played with a piece of wire, twisting it one way and then another. After three hours, he still hadn't thought of a way to get out of debt but he had twisted the wire into an interesting shape.

The pins available in Hunt's day were straight pins, with a tiny head at one end and a sharp point at the other. They were used for sewing, to hold hems and seams in place until the tailor or the seamstress got around to sewing the fabric together. Pricking your fingers on a straight pin was an occupational hazard—but that was about to change.

Based on his original piece of twisted wire, Hunt made a pin that bent in two, had a tiny spring at the bottom and a little clasp or catching device at the top that locked the sharp point in place. If a button popped off or a small tear damaged one's clothes, Hunt's safety pin would provide a temporary, stab-proof fix until the button could be sewn back on or the rip could be repaired.

From Hunt's point of view, his invention was a godsend—he still owed that $15 but he could rely on his fallback plan, in which he had promised an investor to sell him his next invention for $400. And that is exactly what Walter Hunt did. He gave away the rights to his safety pin for $400, paid off the $15 debt, and watched during the following years as his safety pin generated millions of dollars in sales.

Devotees of the punk movement adopted the safety pin as an identifying icon

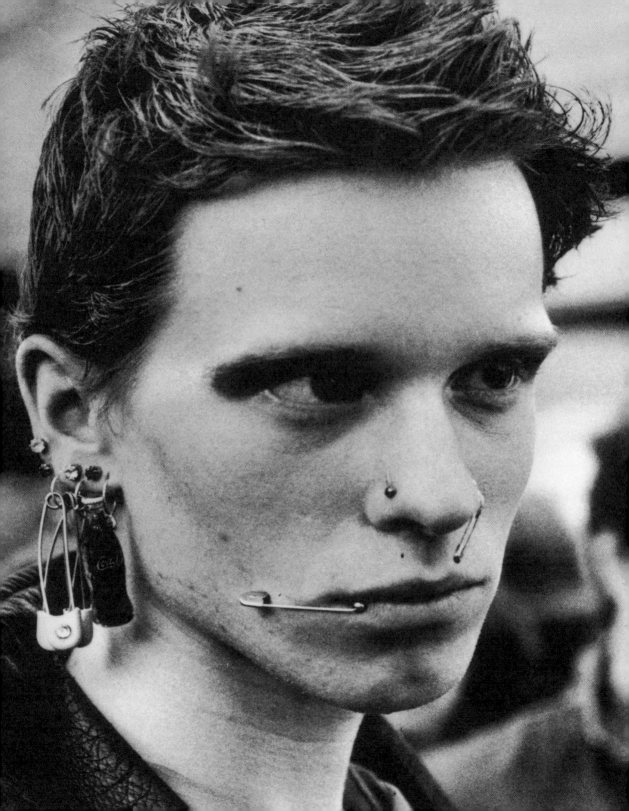

MILKING MACHINE

Milking a cow by hand twice a day, every day, was part of the drudgery of farm life. For dairy farmers who had dozens of cows the routine must have been debilitating. No wonder, then, that the nineteenth century saw an explosion of gadgets and devices to collect milk quickly and efficiently.

One of the earliest attempts at hands-off milking involved inserting tiny tubes into the cows teats. The tubes opened the teats' sphincter muscles and the milk flowed into a collection basin. Naturally, these tubes had to be very small—some were made of silver or carved ivory, while many farmers found it easier and more economical to use quills plucked from a barnyard goose or turkey. The trouble with these milk catheters is they tended to damage the teats, even wasting milk through chronic dribbling.

Vacuum milkers proliferated in the 1850s. One variety was a large rubber bag that fit over the entire udder. Another rubber variety fit over the udder, but had holes through which the teats descended; a hand crank manipulated the udder and the milk flowed out of all the teats at once. Anna Baldwin of New Jersey invented a milking machine that was operated with a hand pump such as she had over her kitchen sink. The problem with all of these machines was they applied too much pressure and injured the cow.

Since the 1920s milking machines mimic hand milking. They still use a vacuum method, with plastic or stainless steel cups slipped over each teat, but the mechanism alternates between gently suctioning the teat and massaging or letting it rest. In this way, the old injuries caused by nineteenth-century milking machines are avoided.

A woman demonstrates an automatic milking machine

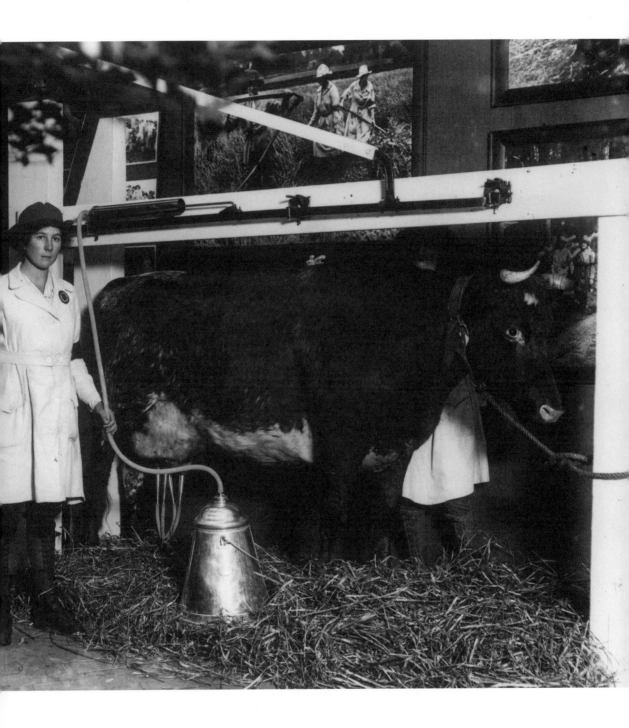

ELEVATOR

The first safety elevator put into operation, located in the Haughwout building in New York

It is said that Archimedes (ca. 287 B.C.–ca. 212 B.C.), the brilliant Greek mathematician and engineer, constructed the first elevator on the island of Sicily, probably to lift wooden beams, stone blocks, and other construction materials. Archimedes' elevator was a wooden platform suspended from heavy ropes that would have run over a sturdy pulley and been lifted either by animal power or human muscle. It was with elevators such as this that the Romans constructed the colosseum and the people of the Middle Ages built the soaring Gothic cathedrals.

During the nineteenth century, when steel was introduced as a new construction material, architects began to erect buildings that were higher than any built before. Like Archimedes' prototype, the elevators of the nineteenth century were regarded as a convenient way to lift heavy loads of freight, but because these elevator cars were raised by a single rope or cable, engineers considered them unsuitable for carrying people—if the cable snapped, the car would plunge down the elevator shaft and everyone on board would be killed, or at the very least badly injured.

It was a Vermont native named Elisha G. Otis (1811–1861) who invented the first safety elevator. In 1852, he mounted a hair-raising, attention-grabbing publicity stunt to demonstrate the reliability of his elevator. At New York's Crystal Palace he was lifted in an open-sided elevator high above the heads of an excited crowd of spectators. At a signal from Otis, a burly assistant with an ax severed the cable. The crowd screamed but the elevator car dropped only a few inches—Otis had invented an automatic spring-operated brake system that was activated when the tension of the cable that raised the car went slack.

In 1857, Otis installed the first safety elevator for passengers at the E. V. Haughwout Building at 488 Broadway at the corner of Broome Street in New York City. Haughwout sold china, silver, chandeliers, and other luxury household items to some of the wealthiest and most distinguished families in America (First Lady Mary Todd Lincoln was among his clients). Installation of the elevator was a smart business move—thrill-seekers who came to Haughwout's emporium to ride the safety elevator often lingered to make a few purchases.

Portrait of Elisha Graves Otis, inventor of the safety elevator

Syringe

A Scottish physician, Dr. Alexander Wood (1817–1884), invented a hollow needle for, as he put it, "the direct application of opiates to the painful points" of his patients' bodies. The opiate Wood referred to was morphine, widely used as a painkiller in the nineteenth century for everything from cancer to childbirth. Morphine was used so often that many patients became addicted, including, tragically, Dr. Wood's own wife.

At about the same time, a French physician, Dr. Charles Gabriel Pravaz (1791–1853), developed an almost identical syringe.

There were two major concerns with the original syringes—the points had to be sharpened regularly and they were almost impossible to completely decontaminate between uses. Because of the reused needles disease and infections were passed from patient to patient.

In 1954, in the wake of Dr. Jonas Salk's landmark discovery of a polio vaccine, Becton, Dickinson and Company churned out one million disposable syringes so American children could be inoculated without fear of infection or contamination.

The next step forward came in 1956 when a New Zealand pharmacist, Colin Murdoch, invented a syringe with a plastic cylinder, thereby reducing the risk of injury from broken glass syringe cylinders.

Finally, in 1974, an American inventor, Phil Brooks, designed a disposable syringe, a revolutionary development that became a genuine blessing in the wake of the AIDS health crisis.

Recently, George Margolin of California invented a Disposable Single Dose Safety Syringe. This completely self-contained device made of cardboard contains a sealed, sterile pouch that the holds the medication and a syringe. By pushing a slider on the outside of the package, the syringe emerges, punctures the skin, delivers the medication, then immediate retracts inside the packaging and is locked inside. It is impossible to use the needle again.

1853

Dr. Jonas Salk innoculates a boy using a syringe

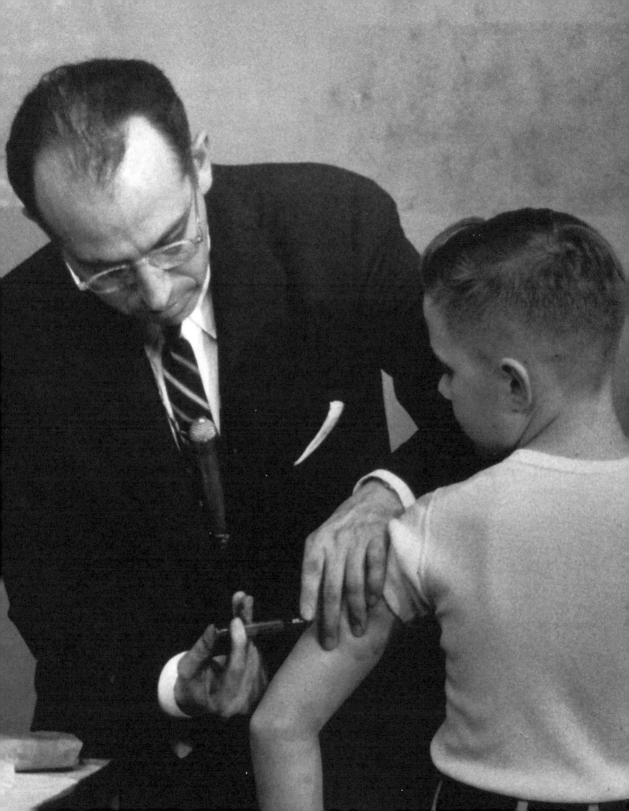

LIFEJACKET

Life vests, lifejackets, life preserver rings, and other personal flotation devices (PFDs) are so commonplace aboard boats and beside swimming pools that it is hard to imagine there was a time when such things were rare. There is evidence that Norwegian sailors went to sea with large blocks of wood or cork they could cling to in case of shipwreck, but such things depended on the strength of the sailor: If his muscles became tired or a large wave dashed the cork from his grip, he would almost certainly drown.

It is said that the first lifejacket was designed by Captain John Ross Ward of the Royal National Lifeboat Institute, founded in 1824 as an organization of rescue crews stationed along the coast of Great Britain and Ireland. Ward's lifejacket was a vest stuffed with cork. It kept a crewman warm in bad weather and also would keep him afloat in an emergency situation. These cork life vests were on the *Titanic*, although like the lifeboats there were not enough for all the passengers. The tragedy of the *Titanic* became the catalyst for legislation that not only required enough lifeboats for all passengers and crew on all seagoing vessels, but also required enough lifejackets.

Good as the cork-filled vest were, they were hard and uncomfortable to wear. This led to the development of lifejackets stuffed with kapok, the fiber of a tropical tree found in Latin America and the Caribbean. The fiber is lightweight, soft, and extremely buoyant; it is stuffed into the vest like down into a pillow.

During World War II, cork and kapok life vests were replaced by inflatable models the troops christened "a Mae West," after the saucy, well-upholstered American comic actress. The inflatable life vest was developed by inventor Andrew Toti (1915–2005) of California who sold his idea to the U.S. War Department in 1936 for only $1,600.

A British sailor wearing an early life preserver

PASTEURIZATION

Portrait of
Louis Pasteur

In 1856, Louis Pasteur (1822–1895) was teaching chemistry at the Faculty of Science in the city of Lille, France, when the father of one of his students, a man named Bigo, asked for the professor's help. Bigo had a beet sugar distillery and he had been running into trouble with his fermentation process: Something was turning his alcohol sour. Pasteur visited the Bigo distillery and began studying samples from the vats. He found that two types of fermentation were going on. The first was the natural process of alcoholic fermentation when yeast was added to the mixture but the second was more mysterious; in some of the vats, Pasteur found lactic acid, the same acid in sour milk. Pasteur identified the lactic acid bacillus, or germ, and suspected that it was airborne. The next questions was how to prevent these contaminants from ruining a batch of beet sugar alcohol.

Since the lactic acid bacillus was a living organism, Pasteur tried heating the alcohol briefly. When he ran another test, he found that the heat had indeed killed the contaminating microorganism. Further experimentation proved that this brief heating process was effective in safeguarding wine, beer, vinegar, and milk from spoilage. In the case of milk, for example, even today, heating it at 163 degrees Fahrenheit for fifteen seconds is enough to kill any bacteria without doing significant damage to the milk's taste or nutritional value. Another advantage is the prolonged shelf life of pasteurized milk—up to two or three weeks.

Since Pasteur's day, pasteurization has extended to fruit juices, ice cream, honey, sports drinks, and even bottled water. Furthermore, the United States Department of Agriculture and the Food Standards Agency in the United Kingdom require milk to undergo the "high temperature/short time" method before it can be sold as pasteurized milk.

Louis Pasteur in his laboratory

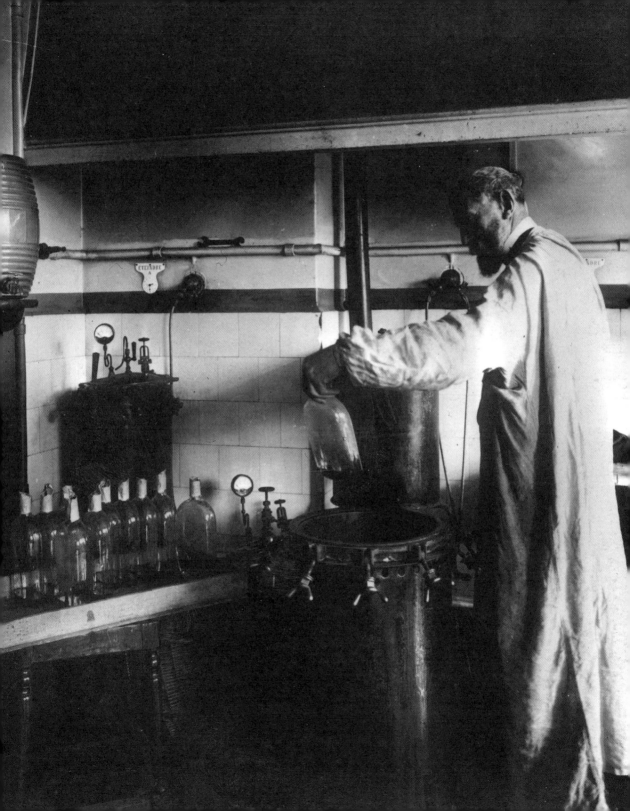

Toilet Paper

If you were a Viking, you used sheep's wool. If you were an Eskimo, you used tundra moss. If you were a peasant, you used hay. And if you were French royalty, you used lace.

The earliest mention of the use of *toilet paper* comes from China: A scholar named Yan Zhitui (531–591) wrote, "Paper on which there are quotations or commentaries from Five Classics or the names of sages, I dare not use for toilet purposes." By 1391, China's Bureau of Imperial Supplies was turning out almost one million sheets a year (each sheet measured two by three feet), for the use of the emperor and his court.

In 1857, Joseph Gayetty of New York produced the first package of what he called "The Therapeutic Paper." Since each sheet contained a generous helping of soothing aloe, claiming that the paper was therapeutic wasn't too much of a stretch. Gayetty sold his bathroom tissue in packages of 500 sheets—and printed across each sheet was Gayetty's name.

In 1890, The Scott Paper Company began manufacturing rolls of toilet paper but, given the reluctance of consumers at this time to even allude to intimate bodily functions, getting the product on to store shelves was a problem. So Scott began to run advertisements designed to educate (or, in fact, terrify) their target audience. "Over 65% of middle-aged men and women suffered from some sort of rectal disease," the Scott toilet paper ad read. Then the ads laid the blame on "harsh toilet tissue [which] may cause serious injury." Scott's toilet paper, on the other hand, was free of all chemicals and was soft and absorbent because it was made with "thirsty fibers." The campaign worked. By 1925, Scott led the world in the production of toilet paper.

The Scott Paper Company frightened potential customers into buying their toilet paper

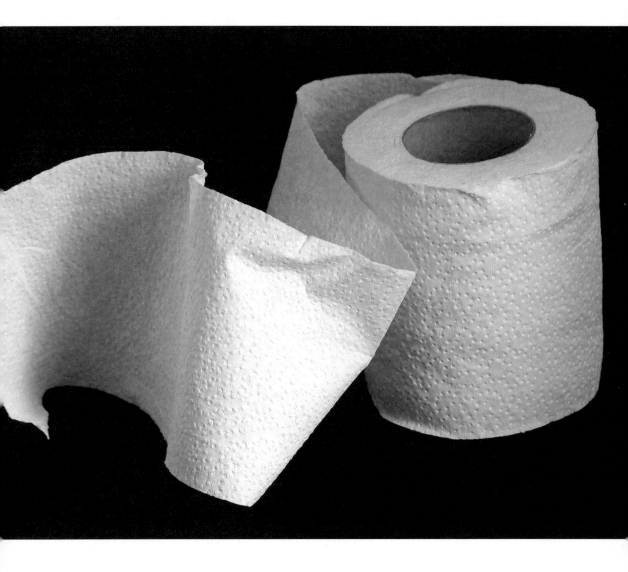

OIL WELL

Since prehistoric times, people had collected oil that seeped up to the surface of the earth, using it as lamp oil, a lubricant for machinery, and even for medicinal purposes. In the 1850s, a New York lawyer, George Bissell, teamed up with a chemist at Yale University, Benjamin Silliman Jr., for the purpose of refining crude oil into a high-quality lamp oil known as kerosene. Rather than try to collect seepage oil, Bissell and Silliman believed that underground lay rich deposits of oil which could be reached by drilling. They selected Titusville, Pennsylvania, as a likely location. The area had many pools of seepage oil, and residents of the neighborhood complained about striking oil when they dug wells looking for fresh water.

To find the oil, Bissel and Silliman, with the financial backing of a New Haven banker, founded the Pennsylvania Rock Oil Company (later the Seneca Oil Company). The company hired Edwin Drake (1819–1880) to dig for oil. Digging proved slow and unprofitable, so Drake switched to drilling and hired a blacksmith who went by the name "Uncle Billy" Smith to run the drilling operation. Using pine boards, they built a derrick, the structure that supports the drilling apparatus, installed a steam engine to power the drill, and went to work. The location they chose was an island on Oil Creek, a tributary of the Allegheny River.

A problem presented itself almost immediately: After the drill had gone down 10 or 20 feet, the walls of the well collapsed. Drake's solution was to create cast iron pipes 10 feet long that were driven into the hole first, then the drill was inserted; as the well grew deeper, more lengths of pipe were fitted together. After weeks of drilling through gravel and then bedrock, Drake and Uncle Billy struck oil on August 27, 1859. They brought up the first oil with a hand pump and poured it into a bathtub.

Drake's discovery set off an oil boom along Oil Creek—the first in American history. Unfortunately, Drake failed to patent his innovative drilling machinery and imitators stole his idea. He sank into poverty. Then, in 1872, the Pennsylvania legislature, in recognition of Drake's contribution to the economy of the state, voted him an annuity of $1,500.

Colonel Drake (in top hat) stands in front of the first oil well

WINDOW SCREENS

Since the 1830s, factories in Connecticut had manufactured wire mesh sieves for use in kitchens. With the outbreak of the Civil War in 1861, when it was no longer possible to export goods to the Confederacy, these manufacturers lost about half of their market. But Connecticut Yankees are nothing if not resourceful, so the manufacturers painted the wire mesh gray and sold it in sheets as window screening.

Before the advent of these first window screens, an open window was an open invitation to flies, mosquitoes, leaves, even birds and squirrels. The window screens still let in fresh air and refreshing breezes but it kept everything else out.

In 1913, screens took an interesting turn. A Czech immigrant to Baltimore, William Oktavec (1885–1956), started a business painting decorative scenes on screens. Initially Oktavec targeted shopkeepers, for whom he painted screens that depicted their merchandise. The screens were so attractive, however, that customers commissioned Oktavec to paint landscapes and other scenes on their window or door screens. These painted window and door screens became known as "Baltimore screens" for the city where they were most popular.

Strange to say—and much to the annoyance of visitors from the United States, Canada, and Australia where window screens are ubiquitous—very few windows in Europe are screened. While there is a certain Old World charm to throwing open the window, leaning out on the sill, and watching the ongoing parade of pedestrians below, the charm is short-lived once flies land on your dinner plate or mosquitoes buzz around your head and keep you awake at night. Even hotels that cater to tourists do not screen their windows.

A woman repairs a window screen

Torpedo

The American Civil War had been raging for barely a year when T. Stoney of Charleston, South Carolina, designed a small steam-driven submarine christened the *David*—a reference to the anemic Confederate Navy pitting itself against the Union Goliath. The *David* was not a true submarine—the smokestack and another pipe to draw fresh air down to the crew both protruded above the water. But what the *David* lost in stealth it made up in a unique offensive weapon. The *David* was armed with a spar torpedo designed by E. C. Singer (nephew of Isaac Singer, inventor of the sewing machine).

Just as the *David* wasn't a true submarine, its torpedo wasn't a true torpedo. It was not shot out of a tube at enemy ships; instead, it was mounted on a pole (or spar) about 40 feet long. Once the *David* had built up a good head of steam, it charged the enemy brandishing its torpedo pole like a knight at a joust. The torpedo was supposed to explode upon contact, and it worked often enough that the Confederate government authorized the construction of a small fleet of 20 *David*-like vessels.

More typical of the time were torpedoes that floated on the surface of the water of a harbor or river, or that were tethered in some way. At the Battle of Mobile Bay in 1864, as Union Admiral David Farragut led his fleet into the bay, his ironclad USS *Tecumseh* struck a tethered torpedo and sank. The rest of the fleet ground to halt, their captains and crew afraid to advance, which led Farragut to exclaim, "Damn the torpedoes! Full speed ahead!"

A self-propelled torpedo was invented by an English engineer, Robert Whitehead, in 1866; he was working at the time for the navy of the Austro-Hungarian Empire. By 1870, he produced a self-propelled torpedo that could travel at a speed of 7 knots and strike a target up to 700 yards away. Navies around the globe watched the development of the new weapon with interest, and finally got a real-life demonstration of its potential in 1891 during Chile's civil war, when a ship of the Chilean navy sank a rebel ship with a single torpedo.

The 1864 battle of Mobile Bay was the first in history to utilize torpedoes

Underground Trains or Subway

A pneumatic subway, the precursor to the underground train

By the 1850s, the congestion of cabs, carriages, freight wagons, trains, trams, and foot traffic in central London was so intense that it took less time to travel by train from London to the seaside resort of Brighton (more than 50 miles away) than to just get across town. As early as the 1830s, Charles Pearson, the City Corporation's Solicitor for London, had urged the government to authorize funds for the construction of underground railroads. Such a transit system would get traffic off of the overcrowded streets, provide easy access to the businesses and markets throughout the city, and at the same time actually increase business as it would no longer be onerous to shop in central London. In spite of these obvious advantages to city life, Parliament always rejected such proposals as too expensive. Not until 1853 did Parliament finally vote to allocate the £150,000 necessary to build the first underground railway.

Constructing such an epic project in the middle of one of the world's busiest and most populous cities proved to be a nightmare. Charles Dickens described it in his novel, *Dombey and Son*: "Houses were knocked down; streets broken through and stopped; deep pits and trenches dug in the ground; enormous heaps of earth and clay thrown up . . . there were a hundred thousand shapes and substances of incompleteness, wildly mingled out of their places, upside down, burrowing in the earth." And in all this chaos and destruction, no thought was given to paying reparation to home and business owners whose property was damaged or destroyed during the construction.

Meanwhile, underground, a kind of hydraulic drill or digger advanced foot by foot through the earth and rock beneath London. As it advanced, huge cast iron segments of pipe—the tunnels for the trains—were set in place and bolted together.

The London Underground opened its doors for business at 6 A.M. on January 10, 1863. Before the day was over, some 38,000 passengers had packed themselves into the cars to ride the underground train, while thousands more were turned away. The first line was only 3.75 miles long; today the Underground has 253 miles of track and carries three million passengers each day.

The world's first underground train line in London shortly before it opened to the public

Dynamite

About 1842, while studying in Paris, Alfred Nobel (1833–1896) learned about nitroglycerine, an extremely volatile liquid explosive. As Nobel's father, Immanuel, manufactured explosives for the construction trade, nitroglycerine became a special focus of the younger Nobel's interest. After completing his studies in Paris, Nobel traveled to Russia where his father was creating explosive devices for the Russian Navy to use in the Crimean War. After the war, Alfred, his brother Emil, and their father opened a laboratory in Stockholm to study nitroglycerine. In 1864, the lab blew up, killing Emil and several others. The calamity did not deter Nobel, however, and in 1866 he discovered that nitroglycerine could be stabilized by mixing it with kieselguhr powder, made from a white chalky sedimentary rock. Nobel named his new explosive dynamite.

Nobel imagined dynamite would be useful in the construction and mining industries but he also marketed it to the military through his connections as owner of Bofors, a cannon factory.

In 1888, a false report that Nobel had died led to the publication of his obituary in a French newspaper under the headline, "The Merchant of Death is Dead." The first line of the obituary read, "Dr. Alfred Nobel, who became rich by finding ways to kill more people faster than ever before, died yesterday." Nobel was horrified by the thought that he would be remembered primarily as an arms manufacturer whose greatest discovery, dynamite, enabled warring nations to kill their enemies more efficiently. He began to think of ways he could improve his reputation and leave a legacy that would benefit the world.

After Nobel's death in 1896 in San Remo, Italy, the executors of his estate reported that in his will, "the merchant of death" had set aside a $9 million bequest for a foundation to award annual cash prizes to individuals who had made the most significant contributions in physics, chemistry, medicine, literature, and peace. The foundation awarded the first Nobel Prizes in 1901.

Alfred Nobel, for whom the eponymous prize is named, was the inventor of dynamite

PERIODIC TABLE OF THE ELEMENTS

Dimitri Mendeleev

John Newlands (1837–1898), a British chemist, first suggested arranging the elements in columns by their atomic mass or weight. Newlands' table, however, was crowded and rigid—he put two elements in each box on his chart, and he left no room for elements which might be discovered in the future. It was a Russian chemist, Dimitri Mendeleev (1834–1907), who designed a better method for displaying the elements, even leaving spaces in his chart for elements which were still unknown but which Mendeleev deduced must turn up eventually.

Mendeleev's Periodic Table listed 66 elements; as of 2007, there are 117 elements on the chart. An element is a pure chemical substance—by definition, no chemical compounds can appear on the Periodic Table. Oxygen, for example, is an element, and so is hydrogen. They appear on the table represented by the letters O and H, respectively. But the mixture of these two elements, H_2O, is a chemical compound (this particular compound is known as water) and so it is not on the chart.

Of the 117 elements, 94 can be found naturally on Earth. The rest have been found elsewhere in the universe among the stars or supernovae, or have been synthesized (created artificially) in laboratories.

The symbols for the elements were first suggested by Jons Jakob Berzelius (1779–1848), a Swedish chemist. The symbols are abbreviations for the Latin names for the elements—Berzelius chose Latin because in his day it was the universal language of science. Today, the symbols of the Periodic Table are understood in every chemistry lab across the globe, irrespective of what language the chemists may speak or even which alphabet they use to write.

1	2	3	4	5	6	7	8	9	10	11	12	13	14	15	16	17	18
																	2 **He** Helium 4.0026
4 **Be** Beryllium 9.0122												5 **B** Boron 10.881	6 **C** Carbon 12.0107	7 **N** Nitrogen 14.0067	8 **O** Oxygen 15.9994	9 **F** Fluorine 18.9984	10 **Ne** Neon 20.1797
12 **Mg** Magnesium 24.305												13 **Al** Aluminum 26.9815	14 **Si** Silicon 28.0855	15 **P** Phosphorus 30.9738	16 **S** Sulfur 32.065	17 **Cl** Chlorine 35.453	18 **Ar** Argon 39.948
20 **Ca** Calcium 40.078	21 **Sc** Scandium 44.9559	22 **Ti** Titanium 47.867	23 **V** Vanadium 50.9415	24 **Cr** Chromium 51.9961	25 **Mn** Manganese 54.938	26 **Fe** Iron 55.845	27 **Co** Cobalt 58.9332	28 **Ni** Nickel 58.6934	29 **Cu** Copper 63.546	30 **Zn** Zinc 65.409		31 **Ga** Gallium 69.723	32 **Ge** Germanium 72.64	33 **As** Arsenic 74.9216	34 **Se** Selenium 78.96	35 **Br** Bromine 79.904	36 **Kr** Krypton 83.798
38 **Sr** Strontium 87.62	39 **Y** Yttrium 88.9059	40 **Zr** Zirconium 91.224	41 **Nb** Niobium 92.9064	42 **Mo** Molybdenum 95.94	43 **Tc** Technetium (98)	44 **Ru** Ruthenium 101.07	45 **Rh** Rhodium 102.9055	46 **Pd** Palladium 106.42	47 **Ag** Silver 107.8682	48 **Cd** Cadmium 112.411		49 **In** Indium 114.818	50 **Sn** Tin 118.71	51 **Sb** Antimony 121.76	52 **Te** Tellurium 127.6	53 **I** Iodine 126.9045	54 **Xe** Xenon 131.293
56 **Ba** Barium 137.327		72 **Hf** Hafnium 178.49	73 **Ta** Tantalum 180.9479	74 **W** Tungsten 183.84	75 **Re** Rhenium 186.207	76 **Os** Osmium 190.23	77 **Ir** Iridium 192.217	78 **Pt** Platinum 195.078	79 **Au** Gold 196.9665	80 **Hg** Mercury 200.59		81 **Tl** Thallium 204.3833	82 **Pb** Lead 207.2	83 **Bi** Bismuth 208.9804	84 **Po** Polonium (209)	85 **At** Astatine (210)	86 **Rn** Radon (222)
88 **Ra** Radium (226)		104 **Rf** Rutherfordium (261)	105 **Db** Dubnium (262)	106 **Sg** Seaborgium (266)	107 **Bh** Bohrium (264)	108 **Hs** Hassium (277)	109 **Mt** Meitnerium (268)	110 **Ds** Darmstadtium (271)	111 **Rg** Roentgenium (272)	112 **Uub** Ununbium (277)							

57 **La** Lanthanum 138.9055	58 **Ce** Cenium 140.116	59 **Pr** Praseodymium 140.9077	60 **Nd** Neodymium 144.24	61 **Pm** Promethium (145)	62 **Sm** Samarium 150.36	63 **Eu** Europium 151.964	64 **Gd** Gadolinium 157.25	65 **Tb** Terbium 158.9253	66 **Dy** Dysprosium 162.5	67 **Ho** Holmium 164.9303	68 **Er** Erbium 167.259	69 **Tm** Thulium 168.9342	70 **Yb** Ytterbium 173.04	71 **Lu** Lutetium 174.967
89 **Ac** Actinium 227.03	90 **Th** Thorium 232.0381	91 **Pa** Protactinium 231.0359	92 **U** Uranium 238.0289	93 **Np** Neptunium (237)	94 **Pu** Plutonium (244)	95 **Am** Americium (243)	96 **Cm** Curium (247)	97 **Bk** Berkelium (247)	98 **Cf** Californium (251)	99 **Es** Einsteinium (252)	100 **Fm** Fermium (257)	101 **Md** Mendelevium (258)	102 **No** Nobelium (259)	103 **Lr** Lawrencium (262)

SQUARE-BOTTOM PAPER BAG

Francis Wolle, a teacher at a Moravian school in Bethlehem, Pennsylvania, invented the first paper bag in 1852. Actually, it was more of a large envelope than a bag, and the narrow bottom failed to achieve what shopkeepers had hoped it would—encourage shoppers to buy more things because now it would be simpler to carry all their purchases home.

The true square-bottom paper bag that truly holds many items was invented by Margaret Knight (1839–1914) in 1868. Originally from York, Maine, she had worked in textile mills from the time she was nine years old. In 1867, or thereabouts, she took a job at the Columbia Paper Bag Company in Springfield, Massachusetts, where, of course, the envelope-style bags were made. Knight had a flare for mechanics and began designing a machine that would cut the paper, fold it into a square shape, and glue the bottom. She had constructed her prototype in wood but to file for a patent she had to have one made of iron. Knight was working on the metal version when the factory received a visitor, a man named Charles Annan, who showed tremendous interest in her and her machine. He studied it so closely that when he left the factory he was able to make a copy and apply for a patent ahead of Knight.

Margaret Knight was not about to let her idea be stolen. She filed suit against Annan charging him with patent interference. During the trial, Annan told the court that it was absurd to believe a woman could comprehend the complexities of a paper bag-making machine. When Knight produced a small mountain of notes, designs, and diary entries describing her work, it became clear to the court that she most certainly did understand the complexities of the machine—and that she had invented it. The judge ruled in Knight's favor.

Typewriter

Remington Arms, the firearms manufacturer, had enjoyed a booming business all through the American Civil War (1861–1865) but after Robert E. Lee surrendered at Appomattox Court House, Virginia, the United States's need for weaponry dropped off. Then, in 1872, an inventor named Christopher Sholes came to Remington with a mechanical writing machine he and some colleagues had created. They needed a manufacturer who had experience producing precision metal components—and Remington certainly had that. The first Remington typewriter was manufactured in 1874; designers at the factory and Sholes continued to make improvements on the design over the next four years.

It was a wonderful device. First, a sheet of paper was scrolled into the machine, held in place on a roller that would advance the sheet upward as the page was covered with line after line of type. When a letter, number, or punctuation key was struck, it swung up a thin metal arm that hit a tape coated with ink. The impact of the metal-on-ink strike left the imprint of the letter or number or punctuation mark on the sheet of paper. At public demonstrations huge crowds turned out to watch a typist's fingers fly across the keys. But when it came time to purchase a typewriter, most of the crowd walked away.

Between 1874 and 1880, Remington sold only 5,000 typewriters. Remington's mistake had been trying to sell typewriters to private individuals; in the nineteenth century (and to some extent even today), there was a widespread belief that private correspondence must be written by hand. Only when Remington approached businesses that generated a great deal of correspondence and other documents did he find a market for his machine. Between 1880 and 1886, Remington sold 50,000 typewriters, almost all of them to companies and corporations.

In the twentieth century, electric typewriters were developed, followed by typewriters that had a correcting tape which lifted characters off the paper. However, the growing popularity of desktop computers in offices and homes in the late 1980s and early 1990s finally made the typewriter obsolete.

A woman typing on a Sholes typewriter

AIR BRAKES

Tractor trailors utilize air brakes as well

By the time he was nineteen years old, George Westinghouse (1846–1914) of New York had fought in the Civil War, dropped out of college, and received his first patent—for a rotary steam engine. A man who was equal parts inventor, manufacturer, and entrepreneur, Westinghouse was a prolific—and wildly successful—founder of new businesses.

Trains, the hottest thing in transportation in the mid-nineteenth century, fascinated Westinghouse. He noticed, however, that the system of brakes trains used was inefficient and dangerous. When a train's engineer wanted to stop, he sent a signal to his brakemen who had to engage a separate set of brakes on each train car. In an emergency, there was no time to bring the train to a stop, and many injuries and fatalities were the result. Westinghouse spent three fruitless years trying to improve a train's brake system; then he hit upon the idea of using compressed air.

Westinghouse built an air compressor that would be installed in the engine cab. From there, compressor pipes would run back to each train car. By opening a valve, the engineer released compressed air into the pipes that would activate the brakes. Once the train was ready to move again the engineer turned another valve that expelled the air from the pipes. Westinghouse received a patent for his air brakes in 1869 before his 23rd birthday and immediately started his first company—the Westinghouse Air Brake Company. He expanded his line to include signal devices and eventually opened factories in Canada and Europe.

Westinghouse went on to start a natural gas company to light homes in Pittsburgh; he built the first electric generator at Niagara Falls; and he electrified the New York, New Haven, and Hartford Railroad line (known to today as Metro North).

By 1900, Westinghouse owned fifteen different companies that employed 50,000 people and were worth collectively about $120 million. He also received 361 patents in his life—the last one arrived just four days before his death.

CAN OPENER

Sixty years after the invention of the tin can, the first can opener finally arrived on the market. Peter Durand, the Englishman who invented the tin can (actually it was a cast iron can plated with tin), was so intent upon finding a new way to preserve food that he forgot to design an easy method for getting the food out again. For years, the "can opener" was a chisel and hammer.

By the late 1850s, a lightweight steel replaced cast iron as the canning medium. In 1858, an American, Ezra Warner of Waterbury, Connecticut, devised a can opener modeled on a bayonet. It had a curved blade that was hammered into the top of the can than moved, in a sawing motion, all the way around the rim. Since this can opener was in essence a huge, sharp, nasty knife, it was more dangerous than the hammer and chisel. It was not uncommon for shoppers to ask the clerk at the grocery store to open the can for them (better he should run the risk of slicing off a finger than a member of the family).

In 1870, William Lyman of West Meridian, Connecticut, developed a can opener similar to the one you likely have in your kitchen. It had a sharp cutting wheel that clipped onto the rim of the can and moved around the edge, cutting the metal top. It was safe, it was easy to operate, and it was small enough to fit in a kitchen drawer.

Electric can openers with a magnet that held the can in place while a motor rotated the can became popular in the 1960s and 1970s, although they are less popular today—no doubt because so many small appliances compete for counter space and electrical outlets in the average kitchen.

Early can opener

BLUE JEANS

In 1853, twenty-four-year-old Levi Strauss (1829–1902), an immigrant from Bavaria, opened a dry goods store in San Francisco to outfit prospectors heading for the gold fields of California. Among his inventory were bolts of heavy cotton cloth. Strauss had been in business—and thriving—for nineteen years when he received a letter from Jacob Davis (1831–1908), a Latvian immigrant tailor in Reno, Nevada, who made work pants using the heavy cotton cloth from Strauss's store. Davis had begun using copper rivets to reinforce the pockets and the fly, which tended to wear out first and tear; these copper rivets made Davis's work pants extremely durable. He had already sold 200 pairs of his riveted pants; now Davis wanted to patent and manufacture them but he needed a business partner. Was Strauss interested?

Strauss knew a good business opportunity when he saw one and teamed up with Davis at once. They received their patent in 1873 for what became known as blue jeans. They ordered high-quality, heavy-duty cotton cloth from the Amoskeag Manufacturing Company in Manchester, New Hampshire. The combination of Amoskeag cotton and Davis's copper rivets made the blue jeans almost indestructible. By the way, all of their original jeans were what is known today as button-fly, or 501® jeans (501 was the lot number in the factory). In time, the partners expanded their line to include jackets and then shirts.

After World War II, with the advent of casual clothing, blue jeans—now known as "Levi's"—made the transition from work clothes to everyday clothes. In the 1950s, Elvis Presley, Marlon Brando, and James Dean were all photographed wearing jeans, which only made them more popular, especially with baby boomers. As the boomers have aged, they have been unwilling to part with their jeans, so Levi Strauss & Co. has expanded their line to fit the expanding body sizes of their most loyal clientele.

Advertisement for Levi Strauss

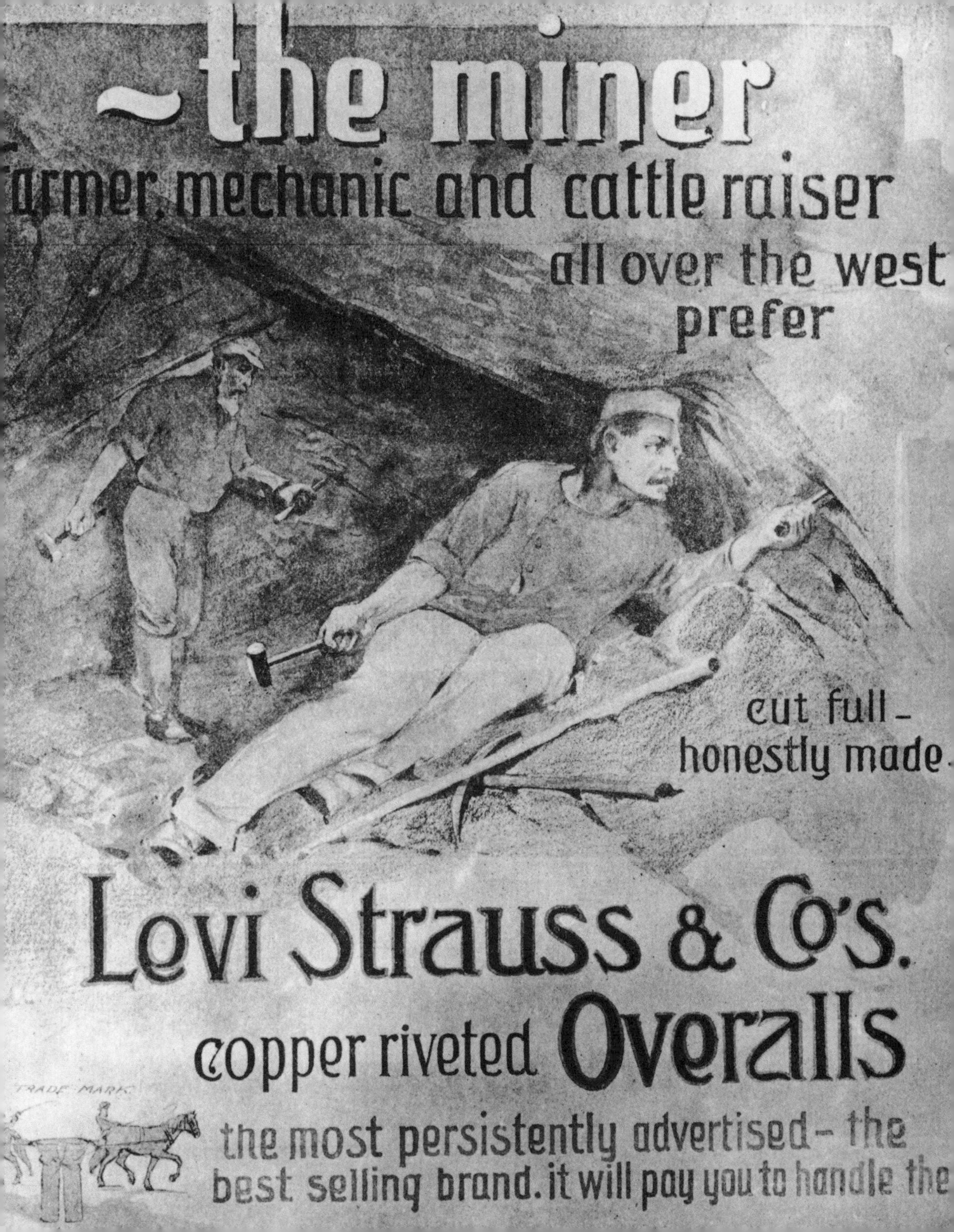
~ the miner
armer, mechanic and cattle raiser
all over the west
prefer
cut full -
honestly made.
Levi Strauss & Co's.
copper riveted Overalls
TRADE MARK
the most persistently advertised - the
best selling brand. it will pay you to handle the

Barbed Wire

The name is much more elegant in French—*fil de fer barbalé*. That is what French inventor Louis Jannin called his brainchild; sadly, he never put his idea into production. It fell to a farmer from DeKalb, Illinois, Joseph F. Glidden (1813–1906), to patent barbed wire and take it to the marketplace in 1874.

In the late nineteenth century, Texas, Kansas, and other wide-open territories of the American West were experiencing a ranching boom. A critical issue for farmers was how to fence in their fields to protect the crops from the vast herds of their neighbors' livestock. Conventional wooden fences were expensive and time-consuming to build—and besides, the cattle broke through them easily. Glidden's barbed wire provided an inexpensive, simple solution to the farmers' problem. All it required were a few fence posts, a spool of barbed wire, and some nails to tack the wire to the posts. The sharp prongs or barbs dissuaded the cattle from trying to stampede through the wire or knock down the fence posts.

With a business partner, Isaac Ellwood, Glidden founded the Barb Fence Company in DeKalb. Their product was an immediate success—they sold 10,000 pounds of barbed wire in 1874 and three million pounds in 1876. When a man named John Warne Gates attempted to capitalize on Glidden's success by selling "bootleg" barbed wire, Glidden, to defend his patent, took Gates to court. The court ruled in Glidden's favor.

Glidden's lawsuit against Gates was miniscule compared with the trouble barbed wire was about to provoke in the West. As fervently as farmers wanted to keep cattle out of their fields, ranchers were just as zealous about keeping the open range open. In other words, they wanted their cattle to be able to wander and graze wherever they liked—as had been customary on the Great Plains before the farmers arrived. The result was a destructive, sometimes violent range war between cattlemen and farmers. This case, too, was decided in court, in favor of the farmers' right to protect their land from the incursions of their neighbors' cattle. It is estimated that within twenty-five years after Joseph Glidden sold his first spool of barbed wire, all privately owned land in the American West had been fenced in.

French farmers repair a barbed wire fence

Internal Combustion Engine

Nikolaus August Otto

Many inventors in the mid-to-late nineteenth century contributed to the development of the internal combustion engine. The man generally accepted as the one responsible for creating a practical, affordable gasoline-powered engine is Nikolaus August Otto (1832–1891).

Otto dropped out of high school to work as a traveling salesman peddling sugar, tea, and kitchen gadgets to general stores on the German side of Germany's borders with Belgium and France. During his travels, he learned about a gasoline-powered engine invented by a Frenchman, Etienne Lenoir. Unfortunately for Lenoir, his design was impractical. It burned up too much gas—100 cubic feet per horsepower per hour. Furthermore, the engine ran so hot that unless the bearings were virtually swimming in oil, the engine would freeze up.

Although he had chosen sales as his profession, Otto was an able mechanic. With a friend, Eugen Langen, he built a two-stroke piston engine that was neither a gas-guzzler nor an engine that ran excessively hot. It won a gold medal at the Paris World's Fair in 1867 but it was not really a practical engine yet. After years of experimentation, Otto introduced to the world the first four-stroke piston cycle internal combustion engine. The piston draws a mixture of gas and air into a cylinder where it is compressed; the compression results in a combustion or explosion inside the cylinder—hence the name "internal combustion engine." Within ten years after its introduction, Otto sold 30,000 of his new engines.

In 1885, Gottlieb Daimler (1834–1900), who had worked with Otto for a time, built a lightweight version of Otto's engine and attached it to a bicycle—this was the first motorcycle. That same year, Karl Benz (1844–1929) mounted one of Otto's engines on a three-wheeled vehicle—this was a prototype of the automobile.

Design for the "Otto" Cycle Gas Engine

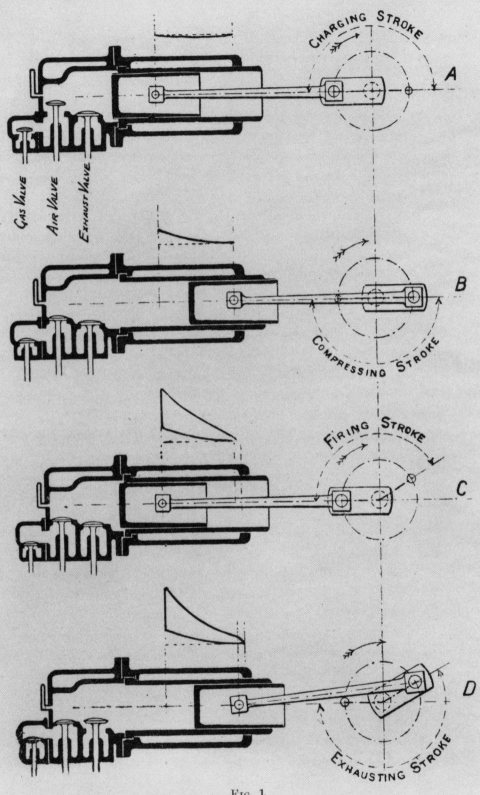

FIG. 1

TELEPHONE

On St. Valentine's Day 1876, Elisha Gray and Alexander Graham Bell (1847–1922) both visited the United States Patent Office. They both applied for a patent for the same invention—the telephone. Who arrived at the office first is a source of controversy but Bell has gone down in history as the inventor of the telephone because he received a patent first.

Bell was a Scot who immigrated first to Canada and then to the United States— consequently, all three countries claim him as one of their own. His primary interest was finding new methods for people who could not hear or speak to communicate. This interest in hearing and sound led to Bell's experiments with sending sounds over telegraph wires. He worked with Thomas Watson (1854–1934), an expert in mechanics and electricity, to find a way to send multiple messages simultaneously over one telegraph wire. During the course of their experiments, Bell and Watson had strung a wire between two reeds such as are used to play musical instruments. Watson casually plucked one of the reeds and the sound carried over the wire to where Bell was working. At that point, the men shifted their research from telegraphy to finding a way to transmit speech over wires.

By February 14, 1876, Bell and Watson had worked out the basic premise of their invention. It was, as the U.S. Patent Office described it, "the method of, and apparatus for, transmitting vocal or other sounds telegraphically . . . by causing electrical undulations, similar in form to the vibrations of the air accompanying the said vocal or other sound." The patent came through on March 7 and, on March 12, Bell placed the first "phone call" when he said, "Mr. Watson, come here! I want you!" Watson, in another room, heard Bell's voice distinctly over the wire. Like the results of Gray and Bell's race to the patent office, the exact words Bell spoke are open to question. Several versions exist but the one quoted here is the one Watso remembered.

The telephone is one of those rare inventions that changed the world, although later in life Bell may have had some misgivings about it. He considered the telephone such an intrusion that he refused to have one installed in his study.

Alexander Graham Bell inaugurates the New York-to-Chicago telephone line

MICROPHONE

The year 1876 marked the 100th anniversary of the signing of the Declaration of Independence and one of the ways Americans celebrated was with the U.S. Centennial Exposition. One of the visitors was Emile Berliner (1851–1929), a twenty-five-year-old immigrant from Germany. At the Bell Company booth, he was fascinated by a demonstration of the telephone but thought the sound quality could be improved. Soon after visiting the exposition, Berliner developed and patented the first microphone. He brought his invention to the executives at the Bell Company who were so impressed they paid him $50,000 for his patent.

Berliner's microphone turned a sound wave into an electrical audio signal. In fact, he was replicating in the telephone how the human ear hears sounds. Sound travels in waves through the air making the air molecules vibrate. Once the vibrating air molecules reach an eardrum, the eardrum vibrates, too, and you hear the sound. Berliner utilized two thin metal plates, known as diaphragms. When a telephone user spoke into the mouthpiece, the sound waves made the metal diaphragms vibrate creating an electrical current that was carried into a cable. This current is the audio signal relayed in telephones and microphones, among other devices.

This basic principle of converting a sound wave into an electrical audio signal has remained constant in the manufacture of microphones. One of the most prominent changes in microphone design is the cordless mic. The handheld model is popular with performers, televangelists, awards show presenters, and amateur singers at karaoke bars. Another type of wireless mic is the lavalier or collar pin. This tiny microphone is clipped onto a collar, a jacket lapel, or some other piece of clothing. A wire from the mic runs to a transmitter, which is usually clipped onto the belt. You can see this type of mic worn by guests and hosts on talk shows, as well as the participants of reality shows. Sound technicians in the entertainment industry tend to dislike wireless mics because they are battery operated, must be monitored constantly, and pick up interference from outside sources.

Emile Berliner with microphones from 1877 and 1927

PHONOGRAPH

In the late nineteenth century, the idea of recording sound occurred to almost no one—except Thomas Edison. Inspired by the new technology of the telephone, Edison believed it would be possible to record and replay sound. His idea was to combine the little vibrating metal plate, or diaphragm, in the mouthpiece of a telephone, with the telegraph stylus that tapped out the dots and dashes on long strips of paper.

He built a simple cylinder covered with tinfoil and operated by a hand crank. Poised on the surface of the tinfoil was a needle that was attached to a diaphragm, which was mounted at the bottom of a big bell-shaped object Edison called the "sound collecting horn." On the evening of December 4, 1877, with his laboratory staff standing around him, Edison turned the crank and recited, "Mary had a little lamb, its fleece was white as snow, and everywhere that Mary went the lamb was sure to go." Then Edison turned the crank again and this time through the sound collecting horn came the sound of his own voice reciting the nursery rhyme.

Here's the science behind what happened: Edison spoke into the horn and his voice caused the diaphragm to vibrate. The vibration of the diaphragm caused the needle to vibrate and scratched the tinfoil. What was scratched into the tinfoil was a sound wave known as an analog signal. For the playback, the process worked in reverse: the needle ran through the grooves scratched into the tinfoil picking up the recorded analog signal, which caused the needle to vibrate, which caused the diaphragm to vibrate, which played back Edison's voice.

A born promoter, Edison went public with his new invention almost immediately, offering demonstrations before large and often skeptical crowds (some people were convinced that Edison's phonograph had to be a hoax). In April 1878, Edison took his invention to Washington, D.C., where he gave a demonstration first before members of the American Academy of Sciences, and then for President Rutherford B. Hayes.

Thomas Edison with a phonograph

REFRIGERATOR

One of the first electric refrigerators ever produced

By the year 1000 B.C., the Chinese were cutting blocks of ice during winter and storing them underground for use to keep food chilled and fresh longer during warm weather. This method remained unchanged into the nineteenth century when many farms and even grand estates had an icehouse where great blocks of ice, packed in hay or sawdust, were kept to preserve food and, of course, make ice cream. Blocks of ice were carried indoors and placed in iceboxes, heavy wooden chests lined inside with zinc or tin. The food was on shelves beside the block of ice, and a pan was placed beneath it to catch the drippings as the ice melted.

All food contains bacteria that, given enough time, will ruin the food and even become toxic (hence, food poisoning). Some foods spoil faster than others; for example, milk left at room temperature for a few hours will go sour. If that same container of milk were put into a refrigerator, the cold temperature will chill the milk, slowing down the action of the bacteria and giving the milk a one- or two-week shelf life.

Carl von Linde (1842–1934) of Germany did not invent a refrigerator but, instead, the chemical that made refrigerators possible. In 1876, he found that a when certain odorless liquefied gas, dimethyl ether, was stored in tanks, the temperature of the tanks was a frosty −13 degrees Fahrenheit (−25 degrees Celsius). Several German breweries were the first industries to adopt Linde's new refrigerant to help them control the fermentation process of the beer and let them store the beer longer.

Two years later, Linde founded the Linde Eismaschinen AG to design and build refrigeration systems for breweries and slaughterhouses. About this time, Linde switched from dimethyl ether to ammonia gas. Other refrigerator manufacturers used methyl chloride or sulfur dioxide—all of these gases are lethal, and in the 1920s accidental leaks of methyl chloride caused several deaths, leading the refrigeration industry to look for safer refrigerants (ultimately they opted for the newly discovered liquid gas, Freon).

By the 1930s, electric powered refrigerators replaced iceboxes in most American kitchens and icehouses vanished from the landscape.

Icebox from 1870's

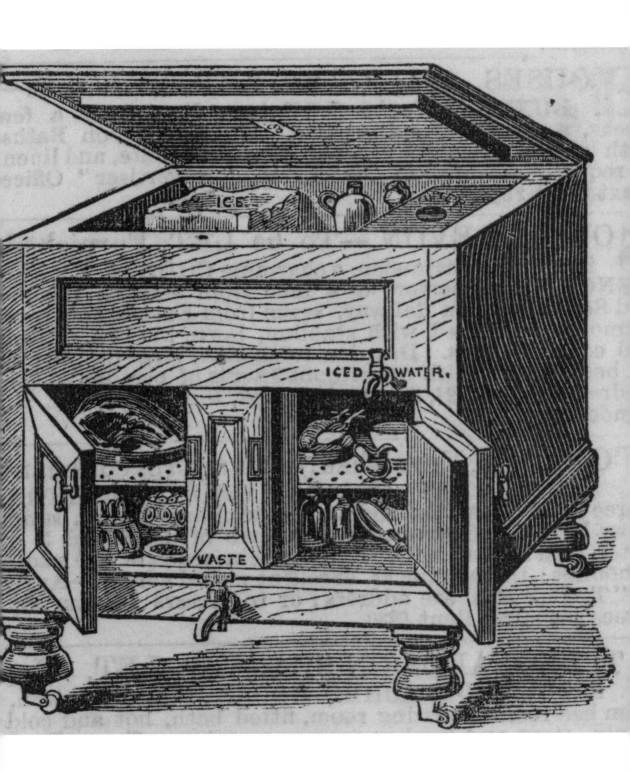

Light Bulb

Diagram of Edison's electric lamp

Every American schoolchild knows that Thomas Edison (1847–1931) invented the light bulb. It was utterly ingenious—the electricity heats the filament inside the bulb until the filament glows. This simple glass bulb with the carbon filament to conduct electricity was simple, practical, economical, and safe (carbon's melting point is 6610 degrees Fahrenheit—and no light bulb has ever gotten that hot). Edison patented his light bulb in 1879. Alas, in 1878 an English inventor, Joseph Swan (1828–1914), had patented a virtually identical light bulb. When he heard about Edison's light bulb, Swan sued. To settle the case, Edison made Swan a partner in the British branch of what became known in Britain as the Edison and Swan United Electric Company.

Now that there were light bulbs, there had to be an energy source to keep them lit. In 1882 in New York City, Edison opened the Pearl Street Power Station, which supplied electricity to 203 homes and businesses. Cleaner and safer than gas, kerosene, whale oil, or candles, electric light was truly dazzling technology. But the light bulbs needed improving: They burned out after 150 hours. The problem was the carbon filament, which Edison made from cotton fibers; in 1886 he changed to a carbon filament made from bamboo—these new bulbs burned for 1,200 hours.

By 1910, three million households and businesses in the United States hummed with electricity.

1878

Thomas Edison in his laboratory in 1915.

CASH REGISTER

The tool room at the National Cash Register Co., circa 1904

Near the train depot in Dayton, Ohio, stood the Pony House, a popular restaurant, saloon, and gambling parlor operated by James Jacob Ritty (1836–1918). The Pony House enjoyed a large clientele of local patrons, traveling salesmen, and even the occasional celebrity (Buffalo Bill Cody is said to have stopped by the Pony House for a few drinks). Ritty's business operated on a cash-only basis, which proved to be an irresistible temptation for some of his employees—they pocketed the money customers had left to pay their bill, rather than placing it in the cash box. In such a busy place, Ritty couldn't watch every transaction of every one of his waiters and bartenders. The problem seemed unsolvable.

Then, in 1878, while on a voyage to Europe, Ritty discovered that the ship had a mechanism that recorded how many times the propeller went around. This gave him the idea for a machine that would record transactions at the Pony House.

Once he returned from his vacation in Europe, Ritty and his brother, John, designed the first cash register. Each transaction was recorded by pressing numerical keys; a locked cash drawer slid open and the money was deposited inside. Once closed, the drawer could not be reopened until another transaction was recorded (except with a special key held by Ritty). Ritty patented his invention as "Ritty's Incorruptible Cashier." He started a factory in Dayton to make cash registers but still operated his saloon. Running two business simultaneously proved too much for Ritty—he sold his cash register factory to a man named Jacob Eckert. A few years later, Eckert sold the cash register business to John Patterson who renamed it the National Cash Register Company—or NCR, as it is known today.

The Pony House was torn down years ago but the huge mahogany bar was preserved and it is still in Dayton, in a restaurant called Jay's.

1879

An early cash register

SHOWER

A shower was only for the wealthy in the ancient world, people who had a staff of slaves to carry in the water, heat it, carry it to the washing room, and then pour it over the master or mistress. This was how King Nebuchadnezzar (605–562 B.C.) of Babylon showered. The Greeks showered in public, standing under a stream of water that flowed from high fountains. But Nebuchadnezzar and the Greeks were not showering as we understand the term, they were getting drenched.

Shower technology stalled after the collapse of the Roman Empire and the destruction of the Roman public bathhouses. In fact, hygiene itself suffered as few people washed anything beyond their hands and face regularly. A major step in a wide acceptance of the shower came in 1879 when the Prussian military commanders ordered communal cold-water showers installed in the soldiers' barracks. The Prussian showers were close to what we find in our bathrooms and locker rooms today, and were light years ahead of a model that was patented that same year by Warren Wasson and Charles Harris, of Carson City, Nevada. The Wasson and Harris shower required the bather to sit on a stool in a bath tub and constantly operate a pump with the feet, thereby sending water up from the bath tub, through a pipe and then out of the shower head.

The hot shower was essentially invented, or at least made possible, by Edwin Rudd in 1898. Rudd built a water tank that heated and stored hot water, ready for whoever stepped under the shower nozzle.

Contemporary shower technology focuses on the quality of the spray with the research and development teams of showerhead manufacturing companies laboring to create a spray that most closely resembles a massage. The other long-sought improvement in shower technology is a device that will keep the water hot even if, somewhere else in the house, someone flushes a toilet.

Edward Poynter's 1884 painting of a bathing woman in ancient Greece

STAPLER

As office jobs increased in the nineteenth century, so did the amount of paper generated. All of those documents had to be sorted and stored properly, which gave rise to a demand for a device that fastened pages together. Some of the first attempts were primitive: folding or crinkling the corners of a document; sewing the pages together with needle and thread; cutting slits through the document and weaving in ribbon or string.

One early mechanical effort for fastening paper was the Improved Eyelet Machine, which punched a circular metal eye through a document. This innovation never became popular because it required a two-step process—once the eyelet had been punched through the front of the document, the document had to be flipped over and the punching process completed on the back. The second was the Brass Paper Fastener, which drilled a hole through the document so a brass fastener could be inserted by hand.

George W. McGill of New York invented the stapler as we know it. It has a reservoir of staples that, one at a time, are driven through two or more pages and, by the same driving action, have their points bent inward to secure the stack of paper. He received a patent for his McGill Single-Stroke Staple Press in 1879. The continuous strip of staples was introduced in 1895 by the Jones Manufacturing Company of Norwalk, Connecticut (later the E. H. Hotchkiss Company).

Interestingly, it was not until 1969, almost a century later, that Joseph A. Foitle of Kansas received a patent for a staple remover.

An early model stapler

Player Piano

Chimes and music boxes were among the first mechanical music-making devices. All other music had to be performed by a person, and even though in the nineteenth century almost every middle-class and upper-class home had a piano, not all pianists were virtuosos. The development of an automated piano was driven by a desire for good music in the home combined with the nineteenth century's passion for mechanical marvels.

The first true player piano was patented by John McTammany of Massachusetts in 1881. It played by reading perforated sheets of paper, the perforations corresponding to individual notes. Alas, McTammany lacked the financial backing and marketing savvy to make a success of his piano. He left the music business and took up a new line of work—inventing voting machines.

The first player piano to make a dent in the American market was the Angelus (Latin for "angel"), which was introduced to American music lovers by an Englishman, Edward H. Leveaux, in 1896. By 1910, player pianos had found their way into many American parlors. To keep up the enthusiasm for their product, player piano manufacturers turned out a steady stream of music rolls (also known as piano rolls) featuring classical music, hymns, dance music, and popular songs of the day. The music roll was attached to spool inside the piano. When the piano was turned on, the spool unrolled the sheet at a regular rate of speed while a tracking device read the perforations and struck the proper notes.

The Depression coupled with the rise of the movies all but wiped out the market for player pianos. People began to look for entertainment outside of their homes. Today, player pianos are the province of collectors and enthusiasts who maintain Web sites and forums where they buy, sell, and trade their instruments and antique music rolls.

Two men compose music for a player piano

ELECTRIC FAN

The earliest fans, probably made from the broad leaves of a plant, were self-propelled, with the overheated person swaying the leaf back and forth to create a bit of a breeze. Pharaohs and others in positions of authority had large fans that were slave-powered. For most of its history, the fan changed very little—whether made of leaves, feathers, paper, silk, wood, ivory, or even metal, it almost always required a person to do the fanning.

Then, in 1882, Dr. Schuyler Skaats Wheeler (1860–1923) of Massachusetts invented a mechanical fan powered by electricity. It had two blades and was small enough to place upon a desk or a table. Wheeler's fan was marketed by the Crocker & Curtis Company, a manufacturer of electric motors. That same year, Philip H. Diehl invented an electric ceiling fan capable of cooling a much larger area than Wheeler's desk model. By the late nineteenth century, it was common to encounter electric fans in offices, shops, and department stores, but aside from the homes of the well-to-do, very few private households had an electric fan.

From the 1890s through the 1920s, electric fans manufactured in the United States were elaborately made. Since they would be positioned prominently in offices and homes, the fans were highly decorative, with brass blades, scrollwork, and brass cages. The cages, however, were not as safe as modern models—often the openings between the bars were so large that an entire hand could fit through which caused many injuries among curious children who couldn't resist touching the rotating blades.

In the 1950s, large box-shaped window fans became available. These came with a mechanism so the homeowner could use the fan either to drive hot air out of the house or to draw cool air from the outside in.

With the advent of air conditioning, however, the market for fans has shrunk.

An electric fan at a switchboard in the early twentieth century

SKYSCRAPER

Modern skyscrapers in Boston

Since the day when the first pharaoh dreamed of the first pyramid, people have sought to display their power and skill by erecting buildings that soared to the heavens. For many centuries, the church steeple or temple tower were the tallest buildings in town. But in the nineteenth century, American tycoons wanted to erected office towers that would dominate the skylines of American cities. The tycoons' timing was excellent because the invention of the electric-powered elevator and steel-frame construction enabled architects to design buildings higher than anything the world had seen before. As it happened, they would also have to devise a new method for laying foundations for such enormous structures, and they would have to take into account wind sheer so a powerful gust would not topple the skyscraper.

The first American skyscraper to use all of these new methods was built in Chicago. After the Great Fire of 1871 in which 18,000 buildings were destroyed, the business owners of the city wanted a new dramatic downtown commercial area. Skyscrapers appealed to the business moguls' egos, and such a skyline would give Chicago a fresh, modern, up-to-date look to rival New York, Boston, and Philadelphia.

Between 1884 and 1885, architect Major William Le Baron Jenney erected on La Salle Street in the Loop of Chicago the Home Insurance Building, a nine-story structure with a basement. The height was nothing to get excited about, even after two more floors were added in 1891. What made Jenney's building a "first" was his method of construction: He used an all-steel frame for the building that would carry the load or weight of the entire structure. Before the Home Insurance Building, the walls of a structure bore the full weight, which limited how high a building could go. At a certain point, the load would be too much and the walls would collapse, taking the entire structure down with them. Once the Home Insurance Building was complete, Jenney's steel frame proliferated across the city, becoming known in the construction trades as the Chicago skeleton.

Sadly, Jenney's building was torn down in 1931.

The Chicago Home Insurance Co. building, the first skyscraper

THE CHICAGO BUILDING OF | THE HOME INSURANCE CO.

OF NEW YORK

AUTOMOBILE

Portrait of
Henry Ford

As with most history-altering inventions, many creative minds contributed to the development of the modern automobile, including a Belgian Jesuit missionary to China, Rev. Ferdinand Verbiest (1623–1688), who built a steam-powered car in about the year 1670. That said, the man generally acknowledged to be the inventor of the automobile is a German engineer, Karl Benz, who, in 1885, manufactured the first automobile for sale to the public. It was powered by a gasoline engine, which was also invented by Benz. He called his vehicle the Motorwagen.

In 1901, Ransom Olds (1864–1950) opened America's first automobile factory in Lansing, Michigan, where he manufactured a car he called the Curved Dash Oldsmobile. The name was derived from the distinctive upward curve of the front of the vehicle that made it look like a classic horse-drawn sleigh. Olds introduced an assembly-line system for producing cars. In 1901, he sold 425 cars; in 1902, he sold 2,500; and in 1903, he sold 4,000.

The automobile that changed America was Henry Ford's Model T, which made its debut in 1908. The Model T had a front-mounted, four-cylinder engine that could be powered by either gasoline or ethanol. It's top speed was 45 miles per hour and it got 13 miles to the gallon in the city and 21 miles to the gallon on the highway. It was a rear-wheel drive car and Ford claimed it had "three speeds." Actually, it only had two—the third "speed" was actually reverse.

During its first month, the Ford factory turned out only eleven cars. To increase production, Ford adopted the assembly-line method Olds was using and he kept making improvements to the method. By 1914, Ford had streamlined his assembly line so effectively that a new Model T could be assembled in 93 minutes. That same year, Ford sold 250,000 cars.

In 1908, the first Model T sold for $825. But every year, as Ford introduced more efficient methods in his factory and garnered higher and higher sales, he lowered the price of his cars. In 1916, Ford offered a basic model automobile for only $365. That year, sales of Model Ts skyrocketed to 472,000.

Benz's three-wheel automobile

DISHWASHER

Josephine Cochrane (1839–1913) was well-to-do; she had a kitchen full of servants to wash her dishes. But the servants were often careless and Cochrane was tired of seeing her fine china nicked and chipped. To ensure that her dinnerware was treated with the proper degree of care, she started washing the dishes herself—and discovered very quickly how tedious that job is. There had to be a better way.

In a shed behind her house, Cochrane built a washer with a rack that held dishes, cups, and bowls in place. The rack lay flat inside a copper boiler. A motor turned the rack and soapy water squirted up, washing the dirty dishes. Cochrane used her prototype in her own kitchen and made copies for her friends. Restaurant and hotel owners were the first to recognize the timesaving (and china-saving) possibilities of the dishwasher; to meet this new demand, Cochrane found the Cochrane Crescent Washing Machine Company. At the World's Fair in Chicago in 1893, she demonstrated her machine to the public and won awards for mechanical design and durability.

Businesses remained the primary market for Cochrane's invention, however. In the late 1890s and early 1900s, most American homes did not have large hot-water heaters—and for one load of dishes, the dishwasher went through a lot of hot water. There were also rumors that the machine didn't clean dishes very well, and that it left a soapy residue, which meant the dishes would have to be washed again by hand anyway.

In the 1950s, about forty years after Cochrane's death, the dishwasher finally began to appear in American kitchens. Home water heaters were now much larger and housewives were eager to own the latest labor-saving appliances. Cochrane's company, by the way, was acquired by KitchenAid, a member of the Whirlpool Corporation.

Contact Lenses

Both Leonardo da Vinci and René Descartes considered the possibility of creating an eyesight-correcting lens that could be worn directly on the cornea of the eye, but such a lens was not made until 1887 by a German physiologist, Adolf E. Fick. He was able to make such a lens thanks to the earlier work of a German glassblower, F. E. Muller, who ground the first glass lens thin enough for a human eye to tolerate. Even so, Fick's lenses were still rather large—18 to 22 mm in diameter—and not very comfortable. After a few hours, wearers removed them. In 1888, another German, August Mueller, refined the contact lenses, making them of lighter glass and shaping them to fit the curve of the cornea. Mueller wore his contact lenses himself to correct his severe nearsightedness (or myopia).

Wearing Mueller's contacts was not easy. First, a dab of cocaine was applied to the eyes as a painkiller. Then the wearer had to submerge his or her face in water (if the lens was not applied underwater, there was a risk of an air bubble forming between the lens and the cornea, which could damage the eye).

An American optometrist, William Feinbloom, introduced plastic contact lenses in 1936, which were popularized in the late 1940s by another American, Kevin Tuohy, a California optician.

A Czech chemist, Otto Wichterle, invented soft contact lenses in 1959. The Food and Drug Administration (FDA) approved soft lenses for sale in the United States in 1971. Today, most users of contact lenses wear disposables, which can be removed and thrown out at the end of each day.

1887

A doctor fits a contact lens

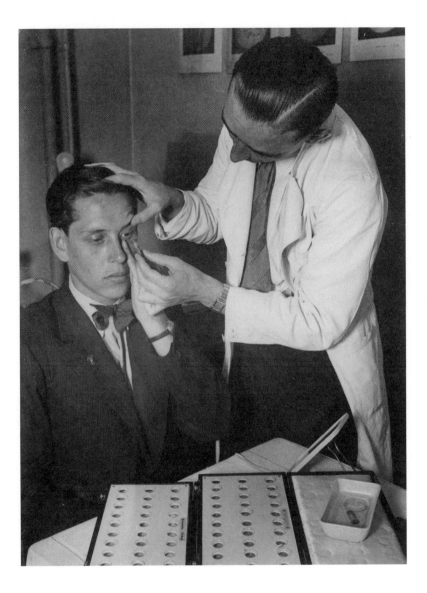

Paper Drinking Straw

The earliest drinking straws were lengths of rye grass which are hollow naturally. In the early Middle Ages, when chalices used at Mass were large and elaborate, priests used straws of gold or silver to drink the consecrated wine (rather than risk spilling it by attempting to drink straight from the cup).

In 1888, Marvin Stone, who made paper novelties, wrapped a thin strip of paper in a spiral pattern around a pencil, glued the edges together, and produced the first paper drinking straw. His prototype had a serious problem—it become soggy and unusable within minutes of being immersed in the beverage. Stone solved the problem by coating his straw with waxy paraffin. Stone's invention was a hit with soda fountain customers across America.

In the 1930s, Joseph B. Friedman (1900–1982), an inventor who also dabbled in real estate, was sitting at his brother's ice cream parlor, the Varsity Sweet Shop in San Francisco, when he had an idea for a new type of paper straw. He inserted a screw down the tube, then wrapped dental floss around the exterior of the straw, following the screw's threads. When he removed the screw and the dental floss, the straw was corrugated, and it could bend without tearing or compressing. In 1937, Friedman patented his flexible drinking straw.

Plastic drinking straws in corkscrew or elaborate knot-like shapes—better known as crazy straws—became popular in the 1960s and are still on the market. While paper straws can still be found on store shelves, the most common straws today are made of plastic and have the flexible corrugated top.

A fashionable women in the early twentieth century drinks from a paper straw

PNEUMATIC TIRES

In about 1887, a Scottish veterinarian, John Boyd Dunlop (1840–1921), presented his young son with a bicycle. The boy was delighted but every time he went out riding on the rough country roads and cobblestones streets, he came home with a headache. His father realized a good part of the problem was the solid rubber wheels of the bicycle—they had no give, they registered every jolt. Dunlop's solution was a rubber tire filled with air—also known as an inflatable pneumatic tire. Because air pressure swelling out in all directions keeps a pneumatic tire solid, that same air acts as a cushion or shock absorber.

Interestingly, there is a preface to this invention: In 1845, the year after Charles Goodyear invented vulcanized rubber, Robert W. Thomson (1822–1873) invented a vulcanized rubber pneumatic tire but it was too expensive to produce and vanished from the market. Dunlop's tire came out at a time when more cost-effective methods of manufacturing vulcanized rubber were available. He received a patent in 1888 but two years later the British patent office took it back after learning that Thomson's patent predated Dunlop's by forty-three years.

Nonetheless, in 1889, Dunlop opened a factory in Dublin, Ireland, to produce pneumatic tires, followed by a second in Birmingham, England. Bicycles were the all the rage in the late 1880s and early 1890s and the bicycle craze was followed by the introduction of the automobile. By the 1920s, Dunlop's company, Dunlop Tyres, was manufacturing pneumatic automobile tires that could withstand speeds of over 200 miles per hour without blowing up. There were no automobiles at the time that could go that fast but racecars were coming on the scene and they would need Dunlop's durable tires.

John Boyd Dunlop with the first bicycle to utilize his invention, the pneumatic tire

MOTION PICTURES (MOVIES)

Stills from Eadweard Muybridge's motion studies

It was an exciting day for the folks in Orange, New Jersey, when "the Wizard of Menlo Park," Thomas Edison, and serial photographer Eadweard Muybridge, held a news conference to announce that they were going to partner up to make moving pictures. The date was February 27, 1888, just a day or two after Muybridge had dazzled an audience at the Orange Music Hall with his zoopraxiscope.

As the first syllable suggests, Muybridge's invention had to do with animals. Using serial photographs he had made of various animals, Muybridge produced a series of paintings that he projected in quick succession on a screen, giving the audience the impression that they were watching a motion picture. For eight months into their partnership, Edison tried to work with Muybridge's zoopraxiscope model but ultimately he was dissatisfied with it. He wanted this new medium to work like his phonograph—just as the phonograph played the sounds as they had been spoken or sung or played by an orchestra, Edison wanted his motion pictures to be a seamless visual record of action. He broke off his partnership with Muybridge and began experiments to record moving images on photographic film.

On May 20, 1891, Edison's wife, Mina, was hosting representatives from the Federation of Women's Clubs. Mrs. Edison invited her guests to tour her husband's laboratories in Menlo Park. There Edison had a special treat waiting for the ladies: It was a very brief film of one of Edison's employees, W. K. L. Dickson, waving hello. The short film thrilled the ladies and as they spread the word, newspaper reporters flocked to the lab to see Edison's latest invention. For the rest of 1891, Edison created short films starring his employees and athletes brought in from nearby Newark.

The movies that followed in the years after *Dickson's Greeting* (as it's known to film buffs) ran for about four minutes or less. The first two-reel "epic" was released in 1904 by George Melies. It was called *The Impossible Voyage* and it ran for an impressive twenty minutes.

Advertisement for the Vitascope, Edison's motion picture invention

ESCALATOR

Jesse W. Reno (1861–1947) called his invention the Reno Inclined Elevator. It was made of wooden slats mounted on an inclined conveyor belt powered by an electric motor. It was set at a 25-degree angle, and at top speed carried passengers to the summit—7 feet up—at 1.5 miles per hour. Initially, it was an amusement ride, first installed in 1896 at the Old Iron Pier at Coney Island, New York. During its two-week stint at Coney Island, 75,000 thrill-seekers rode Reno's Inclined Elevator.

In 1898, Harrods department store in London purchased a Reno Inclined Elevator to carry shoppers to the store's second level. It was thought to be such an overpowering experience that Harrods hired a porter to stand at the top of the contraption and serve a revivifying glass of brandy to passengers who felt faint after the ride.

The same year that Reno received a patent for his invention, Charles A. Wheeler received a patent for an improved model that featured flat steps rather than wooden slats, thus providing a much more stable ride. Eventually, the rights to manufacture Wheeler's machine came to the Otis Elevator Company of New York. At the Paris Exposition of 1900, Otis demonstrated the new device and called it an *escalator*. The name stuck. In 1901, Gimbels department store in Philadelphia installed an escalator.

Once a novelty item, escalators have become ubiquitous. In 2007, there were more than 30,000 escalators in use in the United States, carrying at least 90 billion riders each year. The longest uninterrupted escalator in the United States can be found at the Wheaton Station of the Washington, D.C., Metro System—it is 230 feet long and takes two minutes and forty-five seconds to carry passengers from top to bottom (or from bottom to top). The world's largest freestanding escalator can be found at the CNN studios in Atlanta; it is eight stories tall and climbs through the open-air lobby.

The Harrod's store in London, where the first escalator was installed

CARBORUNDUM

At age twenty-five, Edward Acheson (1856–1931) was hired by Thomas Edison to work in the great inventor's laboratories in Menlo Park, New Jersey. After three years of working on the installation of Edison's new electric lights, Acheson went off on his own; his plan was to invent an industrial abrasive that would be as tough as a diamond—at the time, the hardest substance in the world. His initial idea was to create a synthetic diamond by heating carbon to the temperature at which a diamond is formed in nature. It didn't work, so Acheson drew upon his experiences with electricity in Edison's lab, mixing carbon with clay, then using electricity to fuse the two together. Hard, shiny crystals were the result. Acheson had made silicon carbide but he thought the crystals were a unique compound derived from carbon and fused alumina from the clay known as corundum. By combining these two terms he coined a new term for his invention—carborundum.

Carborundum wasn't the hardest thing on the planet—the diamond still held first place—but it was close. Certainly it was the hardest synthetic substance available. And Acheson's timing could not have been better: American industry needed a high-quality abrasive that could be used for precision grinding, removing burrs from machine components, and polishing or cleaning metal. Carborundum did all of that easily. From his factory in Niagara Falls, Acheson turned out carborundum-coated grinding and cutting wheels, discs and belts, and even carborundum coated sandpaper.

The notion of a manmade product that was virtually the strongest on earth captured the public's imagination, sometimes in unexpected ways. In 1911, when Edgar Rice Burroughs (of *Tarzan* fame) began his "Princess of Mars" or Barsoom, novels, he described the defensive walls of the cities and fortresses on the planet of Barsoom as built of carborundum.

A carborundum factory

Slot Machine

The very first slot machine was made by a San Francisco mechanic, Charles Fey (1862–1944). Like contemporary slot machines, Fey's prototype had three spinning reels. Painted on them were diamonds, spades, and hearts from playing cards, and an image of the Liberty Bell. Players who got three Liberty Bells in a row hit the jackpot—fifty cents in nickels. The first machines Fey made himself in his workshop and then peddled to bar owners. The deal was simple—Fey and the bar owner split the profits from the machine 50–50.

Almost overnight, the slot machine was an enormous success. Fey, working alone, could not keep up with the demand. Gambling equipment companies tried to buy the rights to Fey's invention but he refused to sell. In 1907, Fey got some competition when Herbert Mills of Chicago began manufacturing his own slot machines but instead of symbols lifted from playing cards, Mills' reels bore pictures of fruit—lemons, cherries, and plums. Pictures aside, both Fey and Mills' machines worked identically. Inside the casing were the three reels with ten pictures painted on each. Once the coin had been inserted in the slot and the lever pulled, the three reels spun independently. There was a payoff if, at the end of the spin, three of the same image lined up.

There had been coin-operated machines of chance before Fey. Most of these were modeled on poker and the prize was a free drink at the bar or a free cigar. Fey's was the first machine that delivered cash prizes. Ten nickels wasn't much money but the rattle of the coins cascading into the tray when a player hit the jackpot proved irresistible and became the trademark sound of a gambling parlor.

Gangster and casino impresario Bugsy Siegal installed the first slot machines at his Las Vegas casino, The Flamingo, in 1940. He imagined they would be a pleasant distraction for the wives and girlfriends of the high rollers who played more serious games of chance. Today, however, slot machines are one of the most profitable investments of any casino.

People line up for an early slot machine

Radio

The Marconi
Wireless School
in New York
circa 1912

You cannot see radio waves, yet they do a lot more than transmit AM and FM radio broadcasts. Cordless phones, cell phones, baby monitors, garage door openers, and satellite communications all depend upon radio waves.

Italian inventor Guglielmo Marconi (1874–1937) sent and received the first radio signal in his hometown, Pontecchio, Italy, in 1895. At the time, he believed radio waves would enable him to create a wireless telegraph. But Marconi found little interest in his discovery in Italy, so he moved to England where he set up a radio station on the Isle of Wight. In 1899, he succeeded in sending a wireless signal from England to France, across the English Channel. In 1901, he sent a wireless signal across the Atlantic Ocean, from Poldu, Cornwall, to St. John's, Newfoundland—a distance of approximately 2,100 miles.

Marconi's wireless telegraphy, as it came to be called (for many years the colloquial term for a radio was the wireless) proved to be especially useful aboard ships that began carrying wireless radio communication systems as early as 1899. An especially dramatic incident that demonstrated the value of the radio occurred in 1909 when, after a collision at sea, the ship with a radio was able to call for help, thereby saving the lives of 1,700 passengers. Two heroes of the 1912 *Titanic* tragedy were the doomed ship's radio operators, Harold Bride and Jack Phillips, who remained at their posts until the last minute, trying desperately to reach a ship close enough to rescue the passengers and crew. Bride survived the wreck; Philips died of hypothermia.

Guglielmo Marconi with a telegraph machine

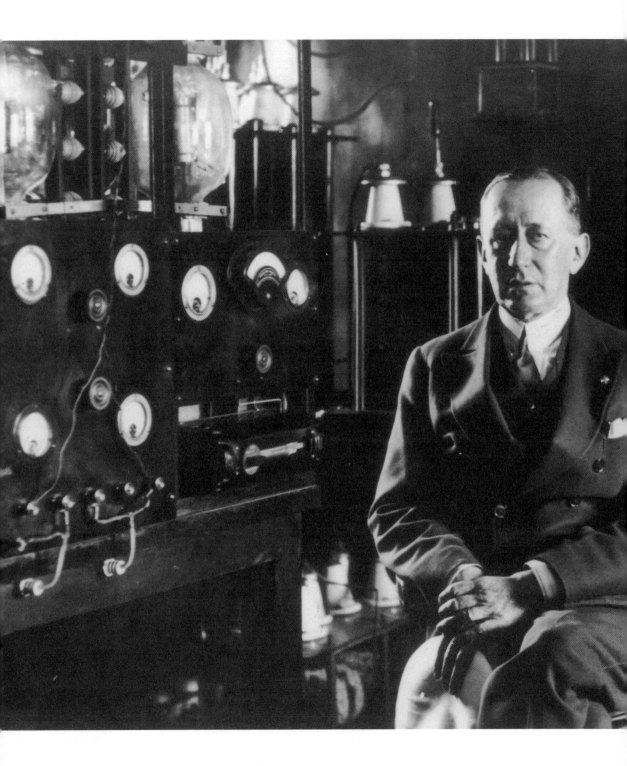

X-Ray

Wilhelm
Roentgen

At the University of Wurzberg in Germany, Wilhelm Conrad Roentgen (1845–1923) was one of the students' least-favorite professors. He had a rapid-fire delivery in the lecture hall that made it difficult for the students to take notes and, although physics came easily to Professor Roentgen, he could not explain the science well to his students. But there was one thing at which Roentgen excelled: He created laboratory experiments for his students that were enlightening and even dramatic.

On November 8, 1895, Roentgen was in his lab studying cathode rays inside a glass cylinder known as a Crooke's tube. It was almost night and the room was dark when he turned on the electron beam; to his surprise, a screen in the lab began to glow with yellowish-green light. He placed objects around the Crooke's tube, yet every time he fired up the electron beam, the screen glowed. One of the objects Roentgen placed against the tube was heavy sheet of cardboard on which a student had painted the letter A with liquid barium platinocyanide—the A was projected on the screen, too. When Roentgen held up a length of pipe to the beam, the image of the bones in his hand were projected on the screen.

Roentgen wrote up his findings for a medical journal, including with his article an X-ray of his wife's hand—each bone and her wedding ring clearly visible. The publication of the report made Roentgen an instant celebrity and the public lined up to see X-rays of the human body.

X-rays are a form of electromagnetic energy with a short wavelength and high level of energy. Roentgen and his colleagues did not realize at the time, however, that radiation was also produced during an X-ray scan, which accounted for the numerous health problems of individuals who had been exposed to too many X-rays.

Nonetheless, the X-ray gave doctors a powerful diagnostic tool—they could examine broken bones, cavities in teeth, and even swallowed objects before resorting to surgery.

A model of Roentgen's X-Ray Tube

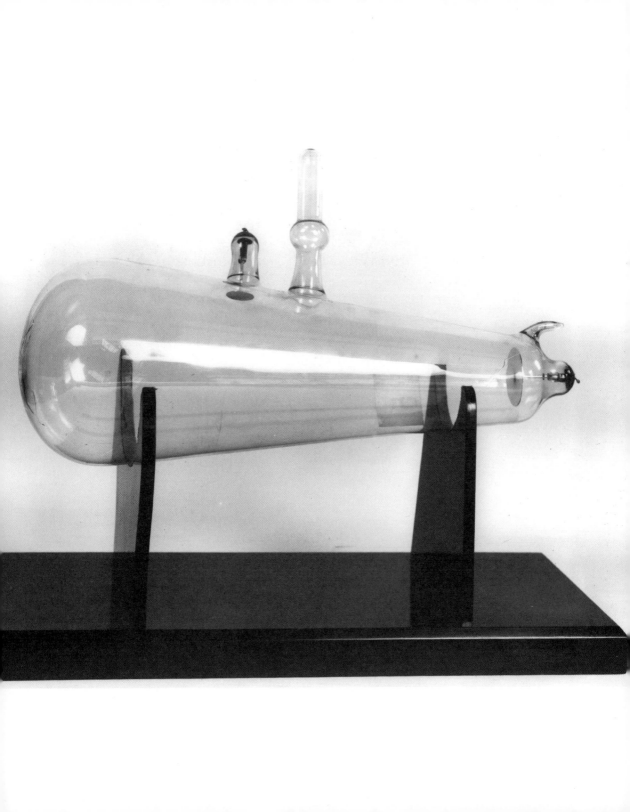

MOUSETRAP

"To build a better mousetrap" has become a colloquial expression meaning that the public is always on the lookout for the next ingenious, labor-saving, life-enhancing invention.

The United States Patent Office has issued patents for more than 4,400 mousetraps but few, if any, have improved upon the design of British inventor James Henry Atkinson, who in 1897 introduced the world to the classic he called "The Little Nipper." It's the mousetrap we all know—little wooden base, tiny tray or pedal that holds the bait, metal arm made tense by a spring, and wire fastener that releases when the unwary mouse takes the bait. The Nipper may be little but it is mighty: Once released, the metal arm slams down on the mouse in 38,000ths of a second, which also explains why it hurts so much when it catches a finger instead of a mouse.

A virtually identical mousetrap was invented in the United States by John Mast of Pennsylvania. He received a patent for it in 1899.

The first live-catch mousetrap on record was built by A. E. "Brock" Kness, a custodian at an Iowa school. In 1924, the school suffered an infestation of mice. Working nights in his garage, Kness invented what he called the "Catch-All Multiple Catch Mousetrap," which could trap several mice without killing them so they could be taken a good distance from the building and released alive into the wild.

There are other types of traps. A bucket trap lures the mouse up a little ramp and into a bucket partially filled with water where the mouse drowns. Glue traps consist of plastic trays coated with an adhesive; the mouse becomes stuck on the glue and dies. There are also electrified mousetraps, which give the mouse a lethal jolt of electricity once it enters the trap. Nonetheless, the snap trap perfected by Atkinson and Mast has never really been improved upon.

1897

The original mousetrap design has not been significantly improved upon

HEARING AID

A man with a hearing aid listens to American politician Henry Cabot Lodge

The earliest hearing aids fall into the category of what became known as *ear trumpets*; made of animal horn or shell or metal and often resembling a funnel or a bugle, these devices were used to amplify sound. The user placed the narrow end of the horn in his or her ear and pointed the broad mouth toward whoever was speaking.

It is often said that Thomas Edison (1847–1931) and Alexander Graham Bell (1847–1922) were involved in the invention of the first electric hearing aid. Edison did contribute to the invention of the microphone, which is an essential component, of course, in a hearing aid. Bell's wife, Mabel Hubbard Bell, was deaf; moved by her condition he founded the American Association to Promote the Teaching of Speech to the Deaf. But Bell did not invent the hearing aid.

The real inventor of the electric hearing aid was Miller Reese Hutchison (1876–1944) of Alabama. In 1899, he received a patent for a hearing aid equipped with a microphone and powered by a battery. He called it the Akoulallion. It was a tabletop model with three earphones so three individuals who were hard of hearing could use it at the same time. But the Akoulallion's cost was prohibitive for most people—it retailed for $400. In 1900, Hutchison brought out a portable hearing aid that sold for $60. There is a science urban legend which claims that Hutchinson went on to invent the intolerably loud klaxon horn in order to drum up more business for his hearing aid company.

Contemporary hearing aids—sometimes called electroacoustic aids—come in a variety of models. One that fits inside or behind the users' ear is among the most common. Other advances in hearing technology include directional microphones, which permit the user to focus on a particular speaker and reduce the background noise in crowded places such as restaurants. And, of course, hearing aids are going wireless.

A early hearing aid; the microphone is hidden in the woman's handbag

Paper Clip

When John Ireland Howe invented his machine to mass-produce straight pins, the New York physician thought he was doing a huge favor for tailors, seamstresses, and everyone else who sewed. And while it is true that straight pins quickly found their way into home sewing kits and tailor and dressmaker's shops, as well as clothing factories, office clerks were also purchasing Dr. Howe's pins to keep pages of documents together. The straight pins worked well, but clerks ran the risk of jabbing their fingers with the sharp end.

Then, in 1899, Johan Vaaler, a Norwegian inventor, designed a metal loop to hold papers together—without risk of bloodshed. At this time, Norway had no patent legislation, so Vaaler applied for and received a patent from Germany. That same year, the ubiquitous double-oval shaped-paper clip that we know today was developed at the Gem Manufacturing Company in England. Impressed by the simplicity and usefulness of the Gem design, William Middlebrook of Waterbury, Connecticut, invented a machine to mass-produce Gem-style paper clips.

There is a story that during the occupation of Norway in World War II, the Nazis forbade Norwegians to wear a likeness of their king or even a pin that displayed his initials. In response, Norwegians began to wear paper clips—a Norwegian invention that holds things together.

The paper clip design has not changed in the many years since its invention

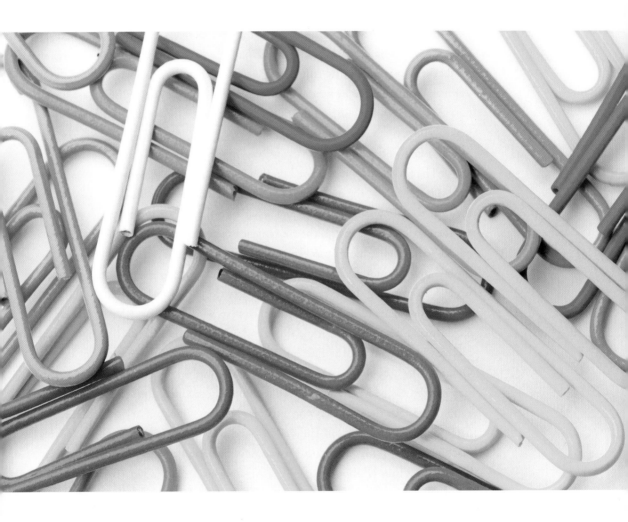

Washing Machine

In the first half of the nineteenth century Catharine Beecher, sister of best-selling author Harriet Beecher Stowe, complained that laundry was "the American house-keeper's hardest problem." She wasn't exaggerating. Everyday women faced a fresh pile of soiled sheets, towels, table linen, clothes, and perhaps diapers, all of which had to be washed by hand. It's been estimated that fifty gallons of water were required to scrub, then boil, and finally rinse a single load of laundry, after which the sopping wet mess would have to be wrung out—by hand—placed in the laundry basket, and lugged outdoors to be hung on the clothesline to dry. Little wonder that many home-makers employed washerwomen.

The first electric washing machine was introduced in the United States around 1900. The precise date it appeared on the market, as well as the name of the inventor, is unknown, but the machine is exactly what housekeepers had been waiting for: they put the dirty laundry, soap, and water into the tub of the machine, turned it on, then walked away. For all the convenience there was a serious design flaw in this model—the electric motor was exposed, mounted on the side of the tub. If the water sloshed over the side it could short out the machine; if anyone touched the motor while it was running she or he might get a nasty electrical shock.

Improvements came at a steady rate, however. The motor was encased in a protective metal housing. The original method, which featured an oscillating tub was replaced by an agitator, the first one being marketed by Beatty Brothers of Fergus, Ontario. The U.S. company that brought agitator washing machines onto the American market was Maytag. In the 1930s manufacturers added timers to the machines, and in the 1950s they added a wonderful innovation—the spin dry cycle that eliminated the need to wring out laundry by hand or by passing it through a wringing machine. In 1957 GE introduced a washing machine with new features that have been standard on all models of ever since—a gauge to set water temperature, the speed of the agitator, and the speed of the spin cycle. It is the fulfillment of Catharine Beecher's dream.

One of the many mechanical innovations leading to the electric washing machine

AIR CONDITIONING

In the ruins of some of the grandest Roman villas, archaeologists have discovered channels built between the walls. These channels were the world's first air conditioners. During hot weather, wealthy Romans ran water constantly through these channels to cool their homes. It was an ideal solution if you were a Roman more troubled by the heat than the humidity.

Willis Haviland Carrier (1876–1960) invented the first practical air conditioner; he was an inveterate tinkerer who grew up in Angola, New York. In 1901, at age twenty-five, he took the first major step toward inventing air conditioning by designing a system that controlled the level of heat and humidity in a printing plant in Brooklyn (the printers had found that when the temperature and humidity levels rose to a certain point, their colored inks did not print properly). Carrier invented a device that drew in the existing air through a filter, passed it over refrigeration coils and blew the cooled, dehumidified air back out, thereby lowering both the temperature and the humidity in the printing plant. Carrier's air conditioner could lower humidity in the plant to 55 percent, and it came with a gauge so the plant operators could adjust the temperature and humidity levels themselves.

In 1915, Carrier and six partners pooled their money and with $32,600 started the Carrier Engineering Corporation to market their cooling systems to large venues. The federal government hired Carrier to install air conditioning in the White House, the Supreme Court Building, and the U.S. Capitol, home to the House of Representatives and the Senate. Carrier also air conditioned New York's Madison Square Garden. By 1922, Carrier found that if he used a centralized compressor in his air conditioners, he could manufacture smaller units that were suitable for movie theaters, offices, railroad cars, and eventually homes. He introduced his Weathermaker air conditioner for homes in 1928. Unfortunately, the Great Depression in 1929, followed by World War II, derailed Carrier's home air conditioning business.

After World War II, however, window-unit air conditioners appeared on the market. In 1948, manufacturers sold 74,000 window air conditioners; in 1953, sales passed the one million mark.

A burlesque theater in the 1920s advertises its Frigidaire air conditioning

Safety Razor

After King C. Gillette's (1855–1932) family lost their hardware supply business in the Great Chicago Fire of 1871, he took a job as a traveling salesmen. One day Gillette's boss mentioned to him casually that disposable items were becoming big business in America.

At the time, men shaved with a large, sharp straight-edged razor that was as a much a weapon as a grooming tool. One slip of the hand while shaving and a man could give himself a nasty cut (for this reason many men went to the local barber shop every morning to be shaved by a professional).

In 1895, Gillette had an idea for a razor comprised of a small, sharp piece of sheet steel mounted in a small T-shaped handle. The blade would be so inexpensive it could be thrown out once it became dull. It took Gillette six years to find an engineer who understood his idea and could help him make a prototype. Gillette's partner was William Emery Nickerson, a graduate of the Massachusetts Institute of Technology. Nickerson and Gillette's design was simple: a handle with a slightly curved T-shaped mount at the top. The disposable razor blade was laid on the mount. A metal cover that left the edge of the razor exposed was screwed on top. When the razor became dull from repeated use, the user unscrewed the cover, threw away the old blade, and put in a new one.

The two men founded the American Safety Razor Company (now known as Gillette) in 1901, began mass-producing their safety razor in 1903, and received a patent for their invention in 1904. Imitators infringed upon his patent but Gillette was making so much money that he could resolve most of these disputes by buying out his competitors. In his lifetime, he sold tens of billions of little packages of disposable razors but his fortune was wrecked by the stock market crash of 1929.

King C. Gillette, inventor of the safety razor

VACUUM CLEANER

Hubert Cecil Booth (1871–1955) had given himself the day off to visit an exhibition of new technology at London's Empire Music Hall. A small crowd had gathered around an American exhibitor, who was demonstrating a machine that was supposed to blow dust into a collection bin. It didn't work very well. After the demonstration, Booth asked the American why he didn't invent a machine that sucked up dirt instead. The American replied that such a thing was impossible.

Over the next few days, Booth couldn't get the idea of a suction-based cleaning machine out of his head. One night, while having dinner with friends, he suddenly pulled his handkerchief from his pocket, laid it over an armchair, placed his mouth on it and sucked. His impulsive act startled his friends but when Booth saw a mouth-shaped circle of dust on his handkerchief he knew he was on to something good.

In 1901, Booth received a patent for his vacuum cleaner. Initially, he did not envision a vacuum cleaner in every house; instead, he sold them to professional housecleaners. It was just as well: Since few homes at this time had electricity, Booth's machine came with a motor and a pump. The apparatus was so big and heavy, however, that it rarely entered a house—it remained outside on a cart while the cleaning men ran long flexible hoses into whatever building they had come to vacuum.

Booth's big break came in 1902. While making preparations for the coronation of King Edward VII, the caretakers of Westminster Abbey discovered that the carpets beneath the royal thrones were thick with dirt and dust. They brought in one of Booth's vacuum cleaners. The vacuum performed so well the king heard of it and promptly ordered vacuum cleaners for Buckingham Palace and Windsor Castle—both of which were large enough to accommodate Booth's bulky motor.

Early vacuum cleaners were so large, the engine had to be kept outside

FLASHLIGHT

The earliest flashlights were blazing torches. In the long centuries before street lamps, anyone who had to leave the security of home after dark took a torch along. The well-to-do traveled at night with slaves or servants who carried torches for their master and also acted as bodyguards. Lanterns with a candle burning inside came along during the Middle Ages and endured into the early twentieth century, especially in rural areas.

An electric-light flashlight could not be invented until electric light bulbs and electric batteries were invented first. In 1902, a Russian immigrant, Conrad Hubert (1856–1928), invented the first portable flashlight. (Like many other immigrants to America, Hubert also had invented a new name for himself—his given name was Akiba Horowitz.) Hubert started out inventing or selling electrically powered novelty items such as a portable electric fan and a light-up stick pin. He founded the Electrical Novelty & Manufacturing Company and hired a man named David Misell to help him develop new products. Misell had already invented a headlamp that could be mounted on the handlebars of a bicycle; now he and Hubert began working on a handheld electric light.

They agreed upon a tube shape to hold the electric battery, with a brass reflector at one end in which the light bulb was mounted. There would also be an on and off switch. In a burst of inspiration, they traveled around to various New York City police precincts and handed out free flashlights to cops who walked a beat at night. The police took to the flashlights at once. Hubert patented the invention and four years later the National Carbon Company (which evolved into the Eveready company, now Energizer) paid Hubert $200,000 for a half interest in his flashlight manufacturing business.

An example of an early flashlight

Airplane

In Dayton, Ohio, the Wright brothers, Wilbur (1867–1912) and Orville (1871–1948), made their living by designing and repairing bicycles but their passion was something much less earthbound—flying machines. In 1896, the brothers heard news stories about advances in flight. Samuel Langley (1834–1906), secretary of the Smithsonian Institution, built a steam-powered model flying machine that flew for half a mile. Octave Chanute (1832–1910), a Chicago engineer, built and tested hang gliders along the windy southern shore of Lake Michigan. These were significant advances in flight but the Wright brothers felt both Langley and Chanute had fallen short of the mark.

After reading everything they could find on the science of aeronautics, Wilbur and Orville concluded that a successful flying machine must be powered by a motor. Langley's steam engine had moved the technology in that direction; however, controlling the aircraft was still an issue. Hang gliders could get a man off the ground but, once he was airborne, he was often at the mercy of the winds—which accounted for a rash of fatal glider crashes in the late 1890s.

One day in 1899, Wilbur was watching a flock of buzzards overhead when he noticed that when a buzzard wanted to turn to the right or the left it twisted its wings. The brothers set about designing a controlling mechanism that would enable a pilot to steer left and right, increase or decrease his altitude, and bank, or lean, into a turn just like a buzzard in flight (or, for that matter, a bicyclist following a bend in the road). Once they had perfected their controlling mechanism, they built a four-cylinder, twelve-horsepower engine to propel their flying machine.

They traveled from their home in Dayton to the dunes at Kitty Hawk, North Carolina. It was remote, which would spare the brothers the embarrassment of being at the center of a media circus if their test flight failed, and the sand dunes would make for a soft landing if the plane crashed.

On December 17, 1903, with just a local businessman, a boy from the town, and three men from the nearby lifesaving station looking on, Orville made the first successful manned flight—120 feet in 12 seconds at a speed of 6.8 miles per hour and a flying altitude of about 10 feet above the ground.

The first motorized flight by Orville Wright at Kitty Hawk; his brother Wilbur runs alongside

COAT HANGER

On a cold day in November 1903, Albert J. Parkhouse returned to his office from lunch to find there were no more hooks available to hang up his heavy winter coat. It was a common complaint among the employees of the Timberlake & Sons company—not enough coat hooks! Parkhouse was one of many promising inventors Timberlake had hired to help him grow his wire novelties business, so where other men might have been peevish, Parkhouse invented something. Grabbing a length of heavy wire, he twisted it into a long narrow oval shape. Then he took a short piece of wire, curled it around the center of the oval, bent the top part into a hook, and draped his overcoat over the contraption; now he could hang his coat almost anywhere he liked.

During the weeks that followed, Parkhouse tinkered with his coat hanger and he made copies for his fellow employees who came to prefer the coat hanger to the coat hook.

In January 1904, a little more than two months after Parkhouse had invented the coat hanger, the Timberlake company's attorney applied for a patent in which he claimed that the inventor was John B. Timberlake, the owner of the firm. It was typical at the time for a company to claim rights to anything an employee invented while on the job, although most employers rewarded the ingenious employee with something—a raise, a bonus, a promotion, maybe even a percentage of the royalties. There is no hint that Albert Parkhouse got anything from the Timberlake company.

Shortly after Timberlake & Sons patented his idea, Parkhouse quit his job, packed up his family, and moved to Los Angeles. There he started his own wire novelties company. Tragically, he died young, at the age of forty-eight, of a ruptured ulcer.

The Timberlake office building in Jackson, Michigan, where the coat hanger was born was torn down long ago. There is a parking lot on the site now and no mention of the coat hanger or of Albert Parkhouse.

Coat hangers, a relatively new invention, shown in a movie still from the 1920s

K24·64

COLORED CRAYONS

In the 1885, two cousins, Edward Binney and C. Harold Smith, took over the industrial pigment company Binney's father had started years earlier. Their most famous color was red oxide, the shade of dark red paint that you still see on most American barns. In 1900, working from their new mill in Easton, Pennsylvania, the cousins began exploring a new market—American schools. They began with slate pencils for students and then in 1902 developed dustless chalk for teachers, an invention which won them a gold medal at the St. Louis World Exposition.

The company's sales force suggested the next product line to Binney and Smith—colored crayons. At the time, crayons were big, clumsy, and used mainly in industrial settings. Using paraffin wax and the industrial pigments they already had on hand, Binney and Smith introduced the world to Crayola Crayons. About the same thickness as a pencil but an inch or two shorter, the crayons were the perfect size for children's hands. Binney and Smith started with eight different colors in what would become Crayola's trademark yellow-and-green box. A box sold for five cents. The crayons were an overnight sensation among school kids. By the way, the name Crayola comes from the French word *craie*, meaning "chalk," and the English word *oleaginous*, meaning "oily."

Almost half a century passed before the Crayola Company varied its product line. In 1949, it offered a box of 48 colors. In 1958 came the 64-color box—with the built-in crayon sharpener. In 1993, the company introduced the "Big Box" with 96 colors. Today, Crayola sells about two billion crayons in sixty countries across the globe.

According to the company's website (www.crayola.com), America's favorite color is Blue. Cerulean Blue (a shade Crayola introduced in 1990) takes second place and Purple Heart, a shade released in 1997, is third.

A packaging machine sorts Crayola Crayons

WINDSHIELD WIPER

The first automobiles did not come equipped with anything to keep the windshield clear of snow, rain, or condensation. If it became too hard to see, the driver pulled over to the side of the road and wiped off the windshield with a rag. In a heavy rainstorm or snowstorm, drivers had no option other than waiting it out.

In 1903, Mary Anderson (1866–1953) traveled with friends from her home in Alabama to New York City. They arrived just as a winter storm hit the city. While sitting in a trolley car, Anderson watched as the trolley's driver tried—unsuccessfully—to keep his windshield clear of snow and sleet by making repeated stops to wipe the glass. When she returned to her home in Birmingham, she sketched a simple arm on which was mounted a rubber blade. The arm would be attached above the frame of the windshield and connected to a lever inside the car. The driver would operate the wiper manually, turning the lever back and forth to clear the glass of precipitation.

Anderson patented her windshield wiper and tried to sell the rights to a Canadian automobile manufacturer. They turned her down, writing, "We do not consider it to be of such commercial value as would warrant our undertaking its sale." Other manufacturers turned her down because they believed the movement of the wiper would distract drivers and cause car crashes.

After seventeen years, Anderson's patent expired and she did not renew it. This was a major mistake, as the automobile industry was on the verge of a boom. As new models of automobiles were introduced, the windshield wiper became standard equipment. If Mary Anderson had held on to her patent, she would have become a millionaire. Instead, she spent the rest of her life as the owner and manager of an apartment complex in Birmingham, Alabama.

The windshield wiper is now an absolute necessity for motorists

FLY SWATTER

Undoubtedly, the human hand was the first fly swatter, followed quickly by any object that would squash a fly. In Asia and Africa, fly whisks were preferred. Usually made of long strands from a horse's tail attached to a decorative handle of wood, metal, or ivory, these whisks were kept in motion around an individual's head and face to keep the flies at bay. Various tribes in Indonesia and Polynesia, as well as the Maasai tribe in Africa, regarded the fly whisk as an emblem of royal authority. In Thailand, to emphasize the point of the king's unique status, his fly whisk was made from the tail hairs of the rare albino elephant.

In 1827, a fly whisk actually started a war. The French consul to Algeria got into an argument with the Ottoman sultan's governor, Hussein Dey, regarding the money France had promised the Algerian government but never sent. Incensed and insulted, Hussein Dey struck the consul across the face with his fly whisk. France took this assault upon its ambassador as an unforgivable affront and sent in an army to avenge the consul's honor.

The fly swatter as we know it owes its existence to two men from Kansas. The summer of 1905 produced a bumper crop of flies in Kansas, a situation that worried Dr. Samuel Crumbine, a member of the state's board of health—he was certain the flies would spread disease. One day, while at a baseball game in Topeka, he heard the crowd chant, "Swat the ball! Swat the ball!" That gave him an idea: Dr. Crumbine distributed a public health bulletin urging Kansans to "swat the fly."

A schoolteacher, Frank Rose, took Crumbine's advice and developed a tool to assist in the war against flies. He tacked a piece of screen onto the end of a yardstick—it was the first modern fly swatter. The screen was the beauty part; flies can sense the change in air pressure when a solid object such as a rolled up newspaper is being swung at them, which explains why they often get away. Since the screen was porous, however, flies never saw it coming.

Outboard Motor

According to the story, on a hot summer day in 1906, Ole Evinrude, twenty-nine years old, was picnicking with friends on an island in a Wisconsin lake 2.5 miles from shore. Bess Carey, the young woman Evinrude would marry, mentioned she would like some ice cream, so Ole climbed into a boat and rowed 5 miles, round trip, to bring his sweetheart what she wanted. At some point along the way, as he was sweltering in his rowboat, it occurred to him that the trip back and forth to the ice cream parlor would have been much more pleasant, not to mention speedier, if the boat were equipped with a motor.

Ole Evinrude (1877–1934) was five years old when he emigrated with his parents from Norway to Wisconsin. He grew up to be a big, husky kid, ideal (his father thought) for farm work, but Evinrude loved anything mechanical. In the early 1900s, he attempted to design and build his own line of automobiles, but the business never got off the ground. Three years after his row across the lake he and Bess Carey were married, had a baby boy, and Evinrude was looking for his next business venture. He settled on building a detachable motor that could be mounted on the rear of a boat. Such a motor was already available but it was so poorly designed it did not start even 50 percent of the time.

Evinrude's motor consisted of a grooved flywheel around which a cord was wound; by pulling the loose end of the cord the flywheel spun and started the motor. A lever regulated the engine's speed and a simple tiller acted as the "steering wheel" by turning the motor from side to side. Evinrude had built a simple, practical, reliable outboard motor. Bess Evinrude wrote the first advertisement for her husband's invention: "Don't row. Use the Evinrude detachable rowboat motor."

The motor was so successful with fishermen and hunters, that Evinrude has become a synonym for outboard motor.

The original Evinrude outboard motor

Plastic

In the mid-nineteenth century, various scientists and inventors had studied possible uses for synthetic resins, but it was Leo Baekeland, a Belgian chemist, who launched the "Age of Plastics" with his invention of a hard, waterproof, moldable plastic he called Bakelite. He created it by combing phenol (also known as carbolic acid) with formaldehyde; the mixture was heated and then molded into the desired shape.

Baekeland found that his plastic did not conduct electricity and was heat-resistant, which made it an ideal material for electrical insulators, as well as casings for radios and telephones. He produced Bakelite in a rainbow of colors, and used it to manufacturer everything from children's toys to kitchenware to costume jewelry. Baekeland turned out to be a natural entrepreneur. Once, when asked why he had entered the field of plastics, he replied, "To make money."

Since Baekeland's day, a wide variety of plastics have been developed for a host of different uses. Polyvinyl chloride (PVC) is used to make plumbing pipes, gutters, window frames, and even flooring. Polycarbonate is the plastic used for CDs and plastic eyeglass lenses. Inexpensive-to-produce polyethylene is used primary for lightweight shopping bags, while polypropylene is used to make food containers and car bumpers.

One plastic-making company that has transformed what the world uses to prepare and store food is the Tupperware Brands Corporation. The creation of Earl Silas Tupper (1907–1983), Tupperware's airtight seal—the better to preserve food—was what made it a must-have household item. The Tupperware Party was the brainchild of Brownie Wise; in 1950, under the name Tupperware Patio Parties, Wise started selling the products at informal gatherings in people's homes. Through her parties, Wise sold more Tupperware than the stores; Tupper took notice, started an official party division in his company, and offered Wise a job as vice president. Over the next two decades, she hired thousands of women as Tupperware sales reps to run the parties, placing women at the forefront of the plastics revolution.

Filaments of polyethylene are fed into a bath for the production of plastic

Aluminum Foil

Gold and silver foil, also known as gold or silver leaf, were used in the ancient world to gild statues and even as decorative elements on buildings. The foil or leaf was made by beating or rolling the gold and silver into microscopically thin sheets.

Tinfoil was not as decorative as its gold and silver cousins but it was less expensive and it could be used to preserve food (the first metal cans, invented in the late eighteenth and early nineteenth centuries, were plated with tin). Throughout the 1800s, tinfoil was used as the wrapper of choice to keep chocolate and tobacco products fresh.

In the beginning of the 1900s, an interest in aluminum foil as a replacement for tinfoil began to develop. It had several advantages over tin: it did not corrode as easily; its shiny surface made it visually more appealing than tinfoil; and it was found to keep foods fresh longer than tin. The first aluminum foil rolled off the presses of the Dr. Lauber, Neher & Cie., Emmishofen, in Switzerland in 1910, but aluminum foil did not make much of an impact on the American market until the 1920s.

Richard S. Reynolds (1881–1955), the nephew of the tobacco mogul R. J. Reynolds, had a company that produced tin wrappers which he sold to tobacco and candy companies. Reynolds switched from tin to the shinier metal foil in the 1920s when the price of aluminum dropped. Once he started using aluminum, Reynolds discovered it had another benefit—it could be rolled out much thinner than tin, which meant aluminum metal would reduce his overhead in ways tin could not.

As time went by, Reynolds found other uses for aluminum—eye-catching bottle labels; sealed foil bags for food; and even shiny aluminum foil Christmas trees, mounted a rotating base that played "Silent Night" as the tree spun around slowly. The only problem with Reynolds Wrap was it could not be made airtight. Nonetheless, it dominated the food storage market into the 1950s, when Saran Wrap was invented and took the lead.

A soldier in 1940 eats a chocolate bar wrapped in aluminum foil

GEIGER COUNTER

Johannes Wilhelm Geiger (1882–1962), a German nuclear physicist, left his homeland for England in 1906 to study radioactivity with Ernest Rutherford, a brilliant British scientist who would win a Nobel Prize for Chemistry. In 1911, Geiger invented the first device to measure radiation and count the number of alpha particles emitted by radioactive material in real time (hence, the name, "Geiger Counter").

The Geiger Counter is a tube filled with argon gas with a thin wire running up the middle. When an alpha particle enters the tube, it detaches an electron from an argon atom, which sends an enormous number of electrons rushing toward the wire (this action is known as an "avalanche"). All of these electrons collecting around the wire give off a pulse that can be amplified as an electronic signal that is heard by the human ear as a series of clicks. An intense staccato of clicks indicates the presence of a high level of radioactive material.

In 1928, Geiger worked with Walther Müller (1905–1979), still a doctoral student at the time, to improve his Geiger Counter to detect all types of radioactive material. Although the device is still popularly known as the Geiger Counter, it's proper name is the Geiger-Müller Tube. Geiger Counters can detect radioactive contamination, radioactive minerals such as uranium, and even the presence of radon in the basement of one's home.

During the height of the Cold War in the 1960s, particularly after the Cuban Missile Crisis of 1962 which nearly led to nuclear war between the United States and the Soviet Union, the U.S. government issued Civil Defense Geiger Counters to supervisors of public bomb and fallout shelters. In the event of a nuclear attack, these Geiger Counters could be used to measure the level of radioactivity outside to determine if it were safe to leave the shelter.

A worker tests radiation levels with a Geiger Counter after the Three Mile Island accident

NUTCRACKER

Rocks were the world's first nutcrackers. But humans being what they are, it wasn't long before nutcracking tools were being developed. By 200 B.C. the Romans had a heavy bronze nutcracker—two arms with teeth mounted on a pivot that enabled the arms to be squeezed together to break the nut's shell. In medieval Europe, metalsmiths forged iron and brass nutcrackers that resemble pliers. The famous freestanding nutcrackers that inspired the Tchaikovsky ballet were first produced in Germany around 1800: The nut is placed in the figure's mouth and a lever or screw cracks the shell.

The nutcracker most people have in their homes was invented by Henry Marcus Quackenbush (1847–1933) of New York. He was a tireless entrepreneur and inventor. He had started out in life as a gunsmith and one of his first successful inventions was an air pistol. He went on to make an extension ladder but certainly his most successful creation was the solid steel, nickel-plated, spring-jointed nutcracker. Just place the nut between the jaws and squeeze. Quackenbush included in the package four picks to remove the nut from the open shell. Since 1913, 200 million of the Quackenbush model of nutcrackers have been sold (although in the current model the package contains only two picks). Its banner year for sales was 1976 when the company sold 3.6 million nutcrackers.

The nutcracker business did so well that in the 1930s, Quackenbush's factories stopped producing guns altogether and focused on the nutcracker business. With World War II just a few years away, the company's decision seems misguided in retrospect; however, during the war, the Quackenbush company ended up manufacturing shell casings for the armed forces.

1913

STAINLESS STEEL

It is just like a scene out of a Charles Dickens novel: A twelve-year-old boy from a poor family leaves school to take a job in a steel furnace. Yet that is how Harry Brearley's (1871–1948) life began. He washed bottles in the Firth steel mill in Sheffield, England, the town where he was born. After work, he studied at home and eventually was able to complete his education by taking night classes. He remained in Firth's chemical lab, where he developed a reputation for finding innovative solutions to metallurgical problems.

In the 1910s, when the British arms industry was gearing up for a world war, Brearley had been assigned to study the problem of rust and excessive wear in steel rifle barrels. He experimented by adding chromium to steel, then he applied acid to his new type of steel so he could examine it under a microscope—but the acid had no effect on the steel. If acid would not etch the steel, Brearley knew that nothing else would corrode it either.

Brearley's innovation became known as stainless steel for its bright surface finish, which no amount of use diminished. Stainless steel is made by adding at least 10.5 percent of chromium to the iron-carbon mix of traditional steel, along with as much as 8 percent of nickel. In addition to using the new steel for rifle barrels, stainless steel was used for knives, especially tableware (in fact, over the last century, in most homes stainless steel knives, forks, and spoons have replaced silver).

The shiny surface of stainless steel has made it the metal of choice for many other, larger projects. The elegant Art Deco cap of the Chrysler Building in New York is made of stainless steel; the soaring Gateway Arch in St. Louis is also made of stainless steel; so too are the exteriors of the Parliament House in Canberra, Australia, and the Petronas Twin Towers in Kuala Lumpur, Malaysia—not to mention countless roadside diners.

1913

The top of the Chrysler Building in New York is constructed of stainless steel

Turn and Brake Signals

During the silent movie era, Florence Lawrence (1886–1938), originally from Canada, was a star—but hardly a soul knew her name. On-screen credits were rare in the early part of the twentieth century but Lawrence had such a lovely, memorable face that she was known to audiences as "The Biograph Girl" (she worked almost exclusively with the Biograph Studio).

In 1910, Carl Laemmie (the future founder of Universal Pictures) started the IMP Company to make movies and convinced Lawrence to sign with him. He paid her lavishly and instigated a series of publicity stunts to keep her in the public eye. The movie-going public was dazzled by Lawrence's glamour and she became Hollywood's first movie star.

She was also one of the first female stars in Hollywood to have her own automobile. She bought her first before 1913, and became an ardent promoter of the coming automobile age. She also designed instruments that would improve the automobile. Her first idea was a signal that would indicate to other drivers when a car was about to turn right or left and when it was coming to a stop. The right and left turn signals were little wooden arms attached to the rear of the car which the driver operated by pushing buttons. The stop signal, also a wooden arm, was activated when the driver stepped on the brake pedal. Lawrence never patented her inventions; consequently, when automakers began incorporating turn and brake indicators into the design of their cars, Lawrence received no royalties.

Lawrence's movie career did not fare well, either. She was replaced at IMP by Mary Pickford, known as "America's Sweetheart." By 1920, fewer studios were asking her to appear in their movies. In 1938, after suffering burns in a fire, being widowed twice and divorced once, Florence Lawrence committed suicide.

Brake and signal lights on a 1929 British automobile

Armored Combat Tank

A WWI German tank

In 1912, L. E. de Mole, an Australian engineer, submitted to the British War Office his design for an armored combat tank that moved on rotating treads rather than wheels. His idea was rejected and to de Mole's credit he resisted the advice of friends who suggested that he sell his invention to the Germans.

Two years later, World War I broke out and the British realized they needed some type of armored vehicle that could roll over rough terrain, grind barbed wire into the mud, and fire at the enemy while being impervious to any returned fire. Winston Churchill, then First Lord of the Admiralty, urged the construction of tanks, probably based upon de Mole's designs. The first British tank went into action at the Battle of the Somme on September 15, 1916; it did break through the German lines, but it also, tragically, machine-gunned the men of the 9th Norfolk Regiment. The French soon joined the British in building tanks, but the Germans showed little interest in tanks, building only about a dozen before the end of the war.

Under the Nazis, however, tanks became an important part of the Third Reich's war effort. They were not has heavily outfitted with armor and weaponry as British and Soviet tanks but they were very fast and so became essential in the blitzkrieg.

The ancestor of the tank is the siege tower, an armored wooden tower on wheels that was hauled up to the walls of an enemy town or fortress. When the tower was against the wall, a drawbridge was lowered onto the battlements, and the soldiers inside the tower surged across the bridge and onto the walls. These towers were first used by the Assyrians at least as early as the ninth century B.C.

Elephants were also an early form of tanks. Protected by thick leather or metal armor, with several men armed with spears or bows and arrows riding on the elephant's back, these animals struck terror on any battlefield. The Carthaginian general, Hannibal, is famous for bringing war elephants across the Alps into Italy in 218–217 B.C.

A British Mark I Tank, the first of its kind, in 1915

WRISTWATCH

The first portable clock was invented in 1504 in Nuremberg, Germany, by Peter Henlein (1479–1542). It was a large, lumpy thing, nicknamed "the Nuremberg Egg," and it kept lousy time—as the mainspring wound down the timekeeping mechanism slowed down almost to a stop. But it was convenient. Gentlemen liked the fact that they could carry the watch during the working day and then set the watch on a table or nightstand at home where they could see the time without having to wait for the striking of the clock in the local church steeple. Pocket watches became standard equipment for men for the next 400 years.

In the late 1880s, there was a brief fashion among women for small watches worn on the wrist. Men would have nothing to do with anything so feminine. Besides, they argued, such a small mechanism could not possibly keep accurate time. When World War I began in 1915, soldiers and officers found that keeping their pocket watches safe but still accessible was burdensome. Hans Wilsdorf (1881–1960), the founder of the Wilsdorf & Davis, Ltd., watch company, began producing wristwatches (Wilsdorf & Davis became the renowned firm Rolex). Wilsdorf was a German who had learned his art from a Swiss clockmaker; in 1905, he opened a pocket watch shop in London. Wilsdorf's first wristwatches were sturdy, reliable, and the soldiers had to admit that wearing a timepiece on the wrist was convenient. After the war, when the veterans returned home wearing their wristwatches, they convinced the rest of the public that this type of accessory was acceptable for men.

The wristwatch grew in popularity and, by the 1930s, 65 percent of all timepieces sold by the finest watchmakers in Switzerland were wristwatches. High-quality wristwatches have proliferated ever since, and today a fine vintage wristwatch is as likely to be passed down as a family heirloom as is an antique gold pocket watch.

T. E. Lawrence, better known as Lawrence of Arabia, wearing a wristwatch

Drywall

For centuries, the interior walls of homes and other buildings were made of plaster. First lathes, thin narrow strips of wood, were tacked up between the studs or beams of each room. Then a layer of wet plaster was applied over the lathes. When that layer dried, another layer was spread on top of it. And so it went, until the spaces between the studs were filled with plaster. It took about one week to plaster a room. Aside from being time-consuming and messy, even after the job was finished, plaster cracked and chipped easily.

In 1916, the United States Gypsum Company, a manufacturer of construction materials, introduced a 4-foot by 8-foot sheet of dry gypsum pressed between two sheets of heavy-duty paper. They called the product Sheetrock. The building trades, particularly the plasterers, didn't like the new product, complaining that it looked cheap. It was not until World War II, when the government demanded fast, economical construction of military facilities, that Sheetrock or drywall finally found a market. After the war, when millions of people moved out of the cities into the suburbs, drywall once again demonstrated that it was a neat, quick, inexpensive alternative to traditional plaster walls.

Drywall is made by mixing wet gypsum with paper or fiberglass fiber, then pressing the wet mess between two thick paper mats. The sheets are taken to a large heated room, the drying chamber, where they are kept until the wet core of the sheets has dried and set. Today, various additives are included to the wet gypsum mixture to make the drywall fire and mildew resistant. Finished drywall comes in thicknesses of between ¼ and ¾ of inch, and in 4-foot-wide sheets that can be cut to a specified length. In the United States alone, 40 billion square feet of drywall are sold every year.

A man using drywall to create defense housing in 1941

ADHESIVE BANDAGE

Not long after Earle Dickson and Josephine Knight married in 1917, Dickson discovered that his bride was accident-prone. Scarcely a day passed when she didn't cut herself with a kitchen knife. As a gauze and surgical tape salesman for Johnson & Johnson in New Brunswick, New Jersey, Dickson had plenty of samples on hand to tend to Josephine's injuries. Sadly, the gauze was so thick and cumbersome it made Josephine's movements awkward, producing more hurts.

One evening, when Josephine had nicked herself again, Dickson cut a short piece of surgical tape and a little piece of gauze. He pressed the gauze onto the tape, then wrapped the tape around the wound. It protected his wife's injury and kept it clean but it did not get in her way as she worked around the house. Earle Dickson had just invented the first adhesive bandage, better known as the Band-Aid®.

Dickson took his idea to his supervisors at Johnson & Johnson. At first, they didn't see the advantage of Dickson's invention—until he demonstrated how easily he applied such a bandage to himself. Johnson & Johnson brought Band-Aids to the market but they chose to make them in an unwieldy size—18 inches long and 2.5 inches wide. Initial sales were so poor that the company gave away thousands of Band-Aids free to Boy Scout troops. Since boys running around in the woods tend to cut their fingers and scrape their knees, the Band-Aids got lots of use. It was the Boy Scouts who first began to popularize these adhesive bandages.

To meet the new demand, in 1924 Johnson & Johnson started to manufacture Band-Aids in different sizes. In 1939, the company offered sterilized Band-Aids. And in 1958, it switched from fabric tape to vinyl.

Johnson & Johnson made Dickson a vice president and a member of the company's board of directors. At the time of Dickson's death in 1961, the company had sold more than $30 million worth of Band-Aids.

A mid-twentieth century adhesive bandage box

ZIPPER

The first zipper was modeled on the hook-and-eye fasteners often found on shoes in the late nineteenth century. But this primitive closure gadget was awkward to use, as it tended to pop open at unexpected moments; worse still, from the manufacturer's point of view, it could not be mass-produced easily and cheaply.

In 1905, Gideon Sundback emigrated from Sweden to the United States. An amateur inventor, Sundback started imagining ways to improve the device. Since the hook-and-eye system was not trustworthy, he came up with a new method: on one side would be jaws, on the other a beaded edge; a sliding mechanism would draw the two sides together, snapping the beads into the jaws. When the slider was pulled down, it unsnapped the beads and jaws. It worked but the friction of sliding up and down wore out the beads quickly.

Sundback adjusted his design to two rows of metal teeth that interconnected when the slider was raised, and disconnected when the slider was drawn down. Sundback had solved all of the zipper's problems: it was simple, it was strong, it was reliable, it was not bulky, and it was easy and cheap to mass produce little metal teeth attached to narrow strips of fabric.

The zipper got off to a slow start but the outbreak of World War I brought a huge demand for simple, innovative products soldiers could use. The armed forces placed enormous orders for money belts, flight suits, and life preservers that were all secured with zippers. After the war, the Goodrich Company introduced a new style of rubber boot, the Mystic Boot, that closed with a zipper. But the zipper did not replace the buttons at the fly of trousers until 1937.

A word about the name: earlier incarnations of the product had been know as the "Automatic, Continuous Clothing Closure," the "Clasp Locker," and the "C-curity Fastener." It was the president of Goodrich who inadvertently came up with the name that stuck. "What we need is an action word," he told his salesmen, "something that will dramatize the way the thing zips."

WWI ace Eddie Rickenbacker with his zippered flightsuit

Pop-Up Toaster

Toast comes from the Latin word *tostum*, meaning to scorch or burn. For centuries, the two most common ways to toast a slice of bread was spear it with a knife or fork and hold it near an open flame, or lean slices of bread against a metal teepee-like frame that was set over the flame of a kitchen stove. Either way, it was essential to watch the bread carefully since it only took a moment for bread to make the move from perfectly toasted to entirely burned.

Charles Strite was a mechanic at a factory in Stillwater, Minnesota. Every day in the company cafeteria, he saw piles of toast, most of it burned. He thought there had to be a way to get perfectly toasted bread to the table.

Instead of a flame, Strite built a machine that used heated electric coils to toast the bread. The slices were placed in a spring-loaded wire basket. A knob was pushed down to lower the basket into the machine and ignite the heating coils. So far, Strite's machine was no different than toasting with fire but then he added the beauty part—a timer which the cook could set so each piece of toast would pop up at the perfect stage of lightly or darkly toasted doneness.

The pop-up toaster got off to slow start because Strite had invented it at a time when all bread was still sliced by hand. If the cook sliced the bread too thick to fit inside the toaster's wire baskets, the bread couldn't be used in the machine and was essentially wasted. Nine years would pass before a bread-slicing machine appeared on the market. Now that bread could be sliced to a uniform thickness or thinness, sales of Strite's pop-up toaster skyrocketed.

1919

A girl plugs in one of the first pop-up toasters

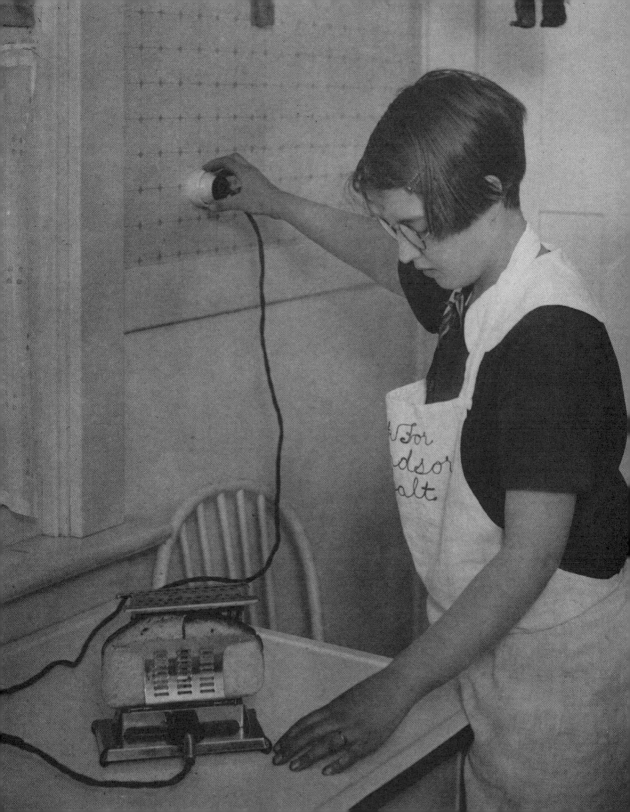

Traffic Light

One of the first traffic lights in Detroit, Michigan

Horse-drawn vehicles were subject to traffic jams, so to keep vehicles and pedestrians moving smoothly and safely through city streets, police officers were stationed at busy intersections to direct traffic. In December 1868, the city of London installed a hand-operated, gas-illuminated traffic light near the Houses of Parliament. A policeman stood beneath it, turning a lever: Traffic that faced the red signal stopped and traffic that faced the green signal were supposed to proceed with caution. But in January 1869, just a few weeks after the traffic signal had been installed, it exploded, injuring the police officer operating it. It would be fifty years before another traffic signal would be attempted.

As the number of automobiles increased, so did traffic jams and accidents. As a Detroit, Michigan, police officer, William L. Potts saw the confusion on the streets firsthand. His idea was to adapt for city intersections the signal lights that were already operating at railroad crossings.

Potts purchased the same red, amber, and green colored glass disks used by railroads, built an electric mechanism to operate the three lights, and placed the invention in a tall rectangular box. Potts's four-way traffic light was placed at the intersection of Woodward and Michigan Avenues in Detroit. It worked so well that, within a year, the city of Detroit had set up fifteen more traffic lights at busy corners. Unfortunately, as a government employee Potts could not patent and profit from his invention.

Claims have been put forward that men other than William Potts deserve to hailed as the inventor of the traffic light, including Lester Wire of Salt Lake City—but Wire's traffic signal had only two lights—red and green. Garrett Morgan of Cleveland, Ohio, is often listed as the inventor of the traffic light, but his invention was not a light—it was a T-shaped sign that read "STOP" or "GO."

Credit for inventing the modern, three-color traffic light belongs to William Potts.

Cotton Swabs

It was bath time for baby in the Gerstenzang house. Leo, an immigrant from Poland, watched as his wife covered the tip of a toothpick with a bit of cotton, then cleaned their baby's ears. Gerstenzang recognized the utility but also the dangers of his wife's invention—one slip and the pointed tip of the toothpick would puncture the baby's eardrum.

It took Gerstenzang a few years to perfect his cotton swab. The little wooden sticks had be harmless, with no sharp ends and no potential to splinter. The cotton swabs had to remain securely on the stick—having one dislodge in a child's ear would be nearly as bad a puncturing an eardrum. Then he had to name his product: Gerstenzang called it "Baby Gays" (in the 1920s, *gay* meant carefree or lighthearted). In 1926, Gerstenzang changed the name of his product to Q-Tips® Baby Gays— the Q stands for Quality.

Ironically, cotton swabs are not effective in removing earwax. The rounded piece of cotton tends to push the wax deeper into the ear canal, actually compressing it and removing very little. Nonetheless, cotton swabs are no longer just for attempting to clear ears. They are used to apply make-up, to take microbiological cultures at doctor's offices and in hospitals, and to apply paint in kids' arts and crafts projects.

As for Gerstenzang, his Q-Tips company made millions. His benefactions ranged from building the science quad at Brandeis University (the quad is named for him) and donating works of art from his private collection to the Museum of Modern Art in New York City.

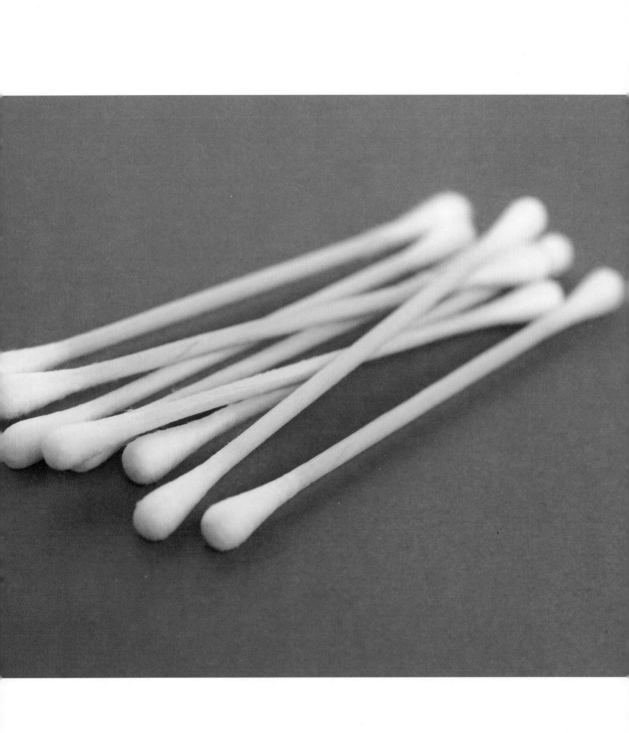

Frozen Food

A Birds Eye frozen food packing facility

In 1912, twenty-six-year-old Clarence Birdseye (1886–1956) of Brooklyn was working as a naturalist near the Arctic Circle. He noticed that in cold weather, Native American fishermen would place their catch on a piece of ice. In frigid weather, the fish froze almost instantaneously—and Birdseye observed that this fast-frozen fish tasted better than fish that had taken a while to freeze. He learned later that in fish that froze very quickly, only tiny ice crystals formed, which did no damage to the cellular structure of the fish; that is why when the fish was thawed, its flavor, color, and texture were as good as when it had been caught.

In 1924, Birdseye was back in the United States and applied for a patent for an invention he called, "Preparing piscatorial products." Once he received his patent he started a company to sell his frozen fish. Unfortunately, his piscatorial freezer didn't freeze the fish fast enough, which gave bacteria an opportunity to interact with the fish, resulting in fish that did not taste good. Birdseye's first frozen food company went belly up.

Eventually, he did figure out how to build a machine that would flash freeze fish as it passed down a conveyor belt. This method worked but the public remained suspicious of frozen food. To help Birdseye market and distribute his product, the Postum Company, a cereal manufacturer, became his partner. After a few years, Birdseye sold his shares to Postum, along with the rights to use his name as a trademark as long as it was split into two words—Birds Eye—for $22 million.

Business really took off after 1930, when Birds Eye took their line of frozen foods—meat, spinach, peas, fruits, and of course fish fillets—to eighteen grocery stores where they offered shoppers free samples. The experiment worked and sales for Birds Eye frozen foods boomed. Today, hundreds of foods can be found in a grocery store's freezer aisle.

Frozen food inventor Clarence Birdseye

ADHESIVE TAPE

By the 1920s, Henry Ford's funereal black Model T was as out of fashion as grandma's corset. Sleek, flashy, two-toned automobiles were all the rage, and while American car manufacturers worked feverishly to meet the demand, they had a tough time overcoming a nearly insurmountable problem—effecting a clean, sharp line at the point where the two colors met. Enter Richard G. Drew (1899–1980), a twenty-six-year-old inventor who had taken a job at 3M in St. Paul, Minnesota. In 1925, after two years of experimentation in the 3M labs, Drew produced the first adhesive tape. It was a 2-inch-wide band of tan colored paper backed with a light coating of pressure-sensitive glue. Drew believed his tape would bring a new level of precision to automobile paint jobs.

Alas, in its first trial run at an automobile plant, Drew's tape proved to be a flop. The painter carefully aligned Drew's tape, pressing it to the car's body, but by the time the painter had picked up his brushes, the tape had fallen off. The trouble was the amount of adhesive on the back of the tape—Drew had applied the glue to the edges of the tape, but not down the middle. "Take this tape back to those Scotch bosses of yours," the frustrated auto painter said, "and tell them to put more adhesive on it!" At the time, the term Scotch as the painter used it meant "stingy."

Back in the lab Drew made the necessary improvements to his tape, and gave it a name—Scotch™ Tape.

In 1930, Drew introduced a variation on his original invention, producing the first transparent cellophane adhesive tape—known at the time as Scotch™ Brand Cellulose Tape.

The self-stick feature of Drew's tape was the beautiful part. Other adhesives on the market in the 1920s were activated by heat or by water. (If you've ever licked the gummed edge of an envelope to seal it, you've encountered a water-activated adhesive.) With Drew's tape, all you had to do was peel it and apply it to where it was needed. Drew's invention was such an instant hit with consumers that its trademarked name entered the American lexicon: Today, adhesive tapes of all brands are known collectively as "scotch tape."

The desire for straight lines on two-toned cars inspired the creation of adhesive tape

Fax Machine

Edouard Belin

The early devices generally referred to as *fax machines* were crude compared with what is available in businesses and homes today. The concept of sending text and images electrically through telegraph wires goes back at least to 1862, to the Italian physicist, Giovanni Caselli. The first truly successful model of the facsimile, or fax, machine was invented in 1925 by a Frenchman, Edouard Belin (1876–1963). To operate Belin's machine, the Belinograph, a person placed a photograph on a cylinder where a bright beam of light scanned it. The machine converted the shades of light and dark in the photo to electrical impulses that could be sent over telegraph wires to a receiver which printed out the image. The Belinograph worked so well that, in 1934, The Associated Press installed the apparatus in its offices so reporters could "wire" photos anywhere in the world.

The first practical fax machine that transmitted letter-size documents over telephone lines was manufactured by Xerox in 1966. It weighed "only" 46 pounds and it took six minutes to send a single page over the wires. In the 1970s, Japanese electronics companies entered the market, producing smaller, faster fax machines. Fax machines today are infinitely faster than the Belinograph and still faster than the models manufactured by Xerox and the Japanese firms. Nonetheless, compared to instant messaging and high-speed Internet connections, the fax is probably the clunkiest piece of office equipment.

The fundamentals of the fax machine have not changed since Belin's day. Contemporary models still operate with a light that scans the document, translates the image into electrical impulses, and then sends them over the wires to the receiver.

A demonstration of a fax machine in 1928

LIQUID-FUELED ROCKET

On October 19, 1899, seventeen-year-old Robert Goddard (1882–1945) climbed into a cherry tree on his family's property to saw off some dead limbs. He had recently finished reading H. G. Wells's novel, *War of the Worlds*, and his imagination was fired by dreams of space travel. As he lopped off the dead branches, he thought, "How wonderful it would be to make some device which had even the possibility of ascending to Mars." Then he wondered "how [a rocket] would look on a small scale, if sent up from the meadow at my feet." For the rest of his life, Goddard observed October 19 as "Anniversary Day," the day he received the inspiration that would give purpose to his life.

In 1914, two years after he had earned his Ph.D. in physics from Clark University, Goddard had received two patents for devices that moved him a bit closer to building a liquid-fueled rocket. In 1920, he began to speak openly of the possibility of a rocket flying to the moon. Friends and colleagues thought Goddard was a bit unhinged, but he had done the mathematics and, on paper at least, it was perfectly feasible to send a rocket to the moon. On a frigid day in March 1926, he launched the first liquid-fueled rocket. Powered by liquid oxygen and gasoline, the rocket rose 41 feet for a flight that lasted 2.5 seconds before diving down into a cabbage field outside Auburn, Massachusetts. In his journal, Goddard recorded, "The first flight with a rocket using liquid propellants was made yesterday at Aunt Effie's farm."

When Charles Lindbergh read of Goddard's successful rocket experiment, he offered to help find the necessary funding for more experimentation. Thanks to Lindbergh, Goddard met the Guggenheim family who financed his research for many years.

Goddard offered his work to the U.S. military establishment but it couldn't see how rockets could be useful in war. Ironically, it was Wernher von Braun, one of the leading research scientists of the Third Reich, who studied Goddard's published work on rocket science and used it to build the V2 rockets that devastated London and other European cities in 1944 and 1945. Tragically, Goddard died in 1945, long before he could see his rockets launched into space.

Inventor Robert Goddard with the first liquid-fueled rocket

AEROSOL SPRAY

An aerosol is made of liquid or solid particles suspended in a gas. It sounds complex until you realize that fog, with its tiny droplets of water suspended in the air, is an aerosol. And smoke, with its tiny particles of ash, is also an aerosol. The idea for aerosol sprays developed from these aerosols that occur in nature.

The nineteenth century saw some advance in the concept of aerosol sprays but the metal containers were too heavy for commercial use. Then, in 1927, a Norwegian engineer, Erik Rotheim, received a patent for a commercially viable aerosol spray can; it was made of aluminum, which made it both lightweight and easy to manufacture. In 1998, the Norwegian government commemorated Rotheim's achievement by issuing a special postage stamp in his (and the spray can's) honor.

Nonetheless, aerosol sprays did not really catch on until World War II. To help protect American soldiers in the Pacific against malaria-bearing mosquitoes, two American researchers working for the U.S. Department of Agriculture, Lyle Goodhue and William Sullivan, produced a small aerosol can for spraying insecticide.

After the war, a twenty-seven-year-old inventor with a strong entrepreneurial streak, Robert Abplanalp, invented a new spray valve that would not clog and could dispense things other than liquids—shaving cream, hair spray, furniture polish, and powdered deodorants. Abplanalp went on to start the Precision Valve Corporation, which produced 1.5 aerosol cans a year and raked in earnings of $100 million annually.

For the aerosol's propellant, Abplanalp, like Goodhue and Sullivan before him, used fluorocarbon gases. Since 1978, when studies revealed that fluorocarbons damaged the ozone layer, manufacturers of aerosol sprays—under orders from the Food and Drug Administration, Environmental Protection Agency, and Consumer Product Safety Commission—have phased out their use of fluorocarbons.

The inventor of aerosol cans, Robert Abplanalp

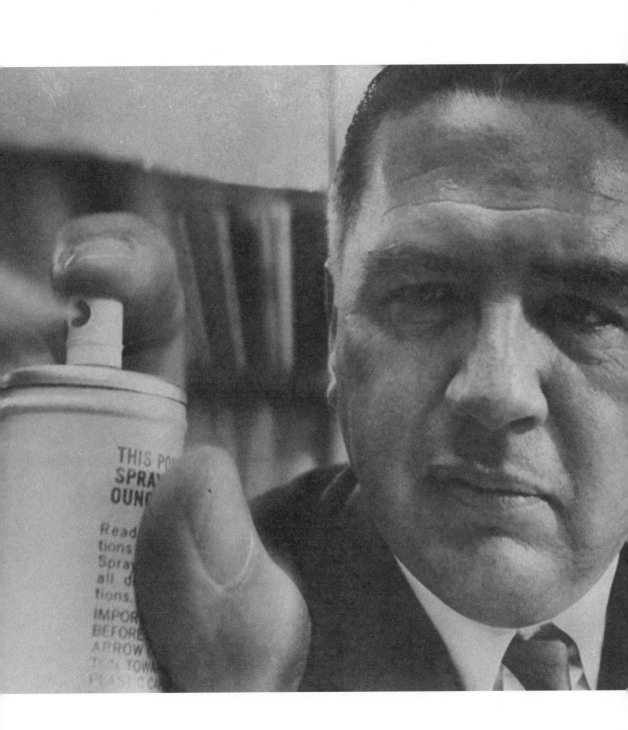

RESPIRATOR

The first man to demonstrate how the human lungs operate was an English chemist and physiologist, John Mayow (1643–1679). He placed a bladder inside a bellows: when he expanded the bellows the bladder filled with air, and when he squeezed the bellows the air was "exhaled" from the bladder.

The first respirator was known as an "iron lung" and was invented in 1927 by Philip Drinker (1893–1977) and Louis A. Shaw (1886–1940), both of whom were professors at the School of Public Health at Harvard University. Their respirator was a large airtight metal box. The patient was laid inside, with his or her head projecting from one end. Connected to the box was a hose and pump that reduced the air pressure inside the sealed chamber, forcing fresh air through the patient's nose and mouth and into the lungs. The first person treated with an iron lung was an eight-year-old girl who suffered from a case of polio that left her chest paralyzed. It was a large, even frightening machine but it saved lives.

Portable, inexpensive respirators were invented by Forrest Bird (1921–), a Massachusetts inventor who had flown for the Army Air Corps during World War II. At that time, planes were being built that could fly at altitudes unimaginable before—but at such heights pilots and their crew could not breathe. This high-flying experience sparked Bird's interest in respiration. During the 1950s, he developed a small respirator in a little green box that in principle was not far removed from Mayow's bellows or Drinker and Shaw's iron lung; to test it, he traveled to hospitals where he asked permission to try his respirator on cardiopulmonary patients who were in critical condition and for whom no other treatment had been successful. In many cases, with Bird's respirator a patient was able to breathe again. By 1970, he had developed a respirator for newborns with breathing problems—before Bird's respirator was in use, the mortality rate among such babies was 70 percent; with Bird's respirator, the mortality rate dropped to less than 10 percent. Since then, at least fifteen different types of respirators are used in hospitals, including a breathing tube that can be inserted directly into the patient's windpipe.

The "iron lung" was the first respirator

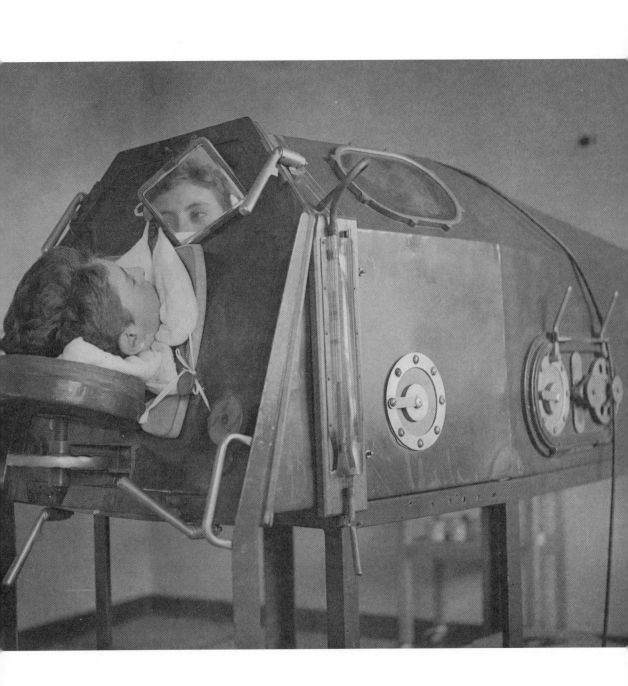

TELEVISION

The debate over who really invented television (TV) can be summed up in the old maxim, "Zworykin had a patent, but Farnsworth had a picture." In the 1920s, Vladimir Kosma Zworykin (1889–1982) was working for Westinghouse, trying to find a way to transmit images the way a telephone or a radio transmitted sounds. At the same time, inventor Philo Taylor Farnsworth (1906–1971), financed by private investors, was trying to achieve the same goal. For a time, it appeared that Zworykin had the upper hand—in 1923 he patented his iconoscope, an early version of a television camera. Unfortunately for Zworykin, his research stalled at the camera, while Farnsworth, on September 7, 1927, transmitted the first images via electronic signals. On that day, TV was born. In 1930 Farnsworth received a patent for his device, which he called "a scanning tube."

Zworykin never managed to even replicate Farnsworth's work and by the early 1930s, his superiors at Westinghouse suggested it was time that he devoted his energy to the research and development of something else.

While television was still in its infancy, RCA, the company that held a virtual monopoly on radio broadcasting, tried to seize control of TV broadcasting, too. (Zworykin, by the way, was now working for RCA.) Farnsworth and RCA went to court to establish whose patent marked the beginning of television. At one point, Farnsworth's high school physics teacher took the stand and testified that when the boy was fourteen years old he had first described to his teacher his ideas for a scanning tube, better known as television. The court ruled in Farnsworth's favor and instructed the executives at RCA that if they wanted to enter the television business, they would have to pay royalties to Philo Farnsworth.

Farnsworth himself appeared on TV only once—in 1957. He was the mystery guest on the television game show, *I've Got A Secret*. The panel failed to guess who Farnsworth was, so he went home with a cash prize of $80.

Philo Farnsworth, the inventor of television, with early TV equipment

Bread-Slicing Machine

"The greatest thing since sliced bread!"—It's an exclamation that has passed into common vocabulary as a tribute to a life-improving invention or idea. And the man who invented the bread-slicing machine was Otto Rohwedder (1880–1960) of Missouri. He was a successful jeweler who owned three jewelry stores; in his spare time, however, Rohwedder tinkered with ideas for new inventions.

In the 1920s, people either baked their bread at home or bought bread from a bakery. Either way, the bread arrived on the table in a solid loaf that had to be sliced by hand. Few people had the eye or the dexterity to cut a loaf into identical slices but Rohwedder believed he could change all that. In 1912, certain that his bread-slicing machine idea would make his fortune, Rohwedder sold his jewelry stores and used the funds to finance years of research and development. In 1917, he had a complete set of blueprints for making his invention, a prototype bread slicer, and a factory in Illinois ready to manufacture the machine.

Tragically, before a single bread slicer rolled off the assembly line, the factory caught fire and burned to the ground, destroying Rohwedder's blueprints and his prototype. To support his family, he went to work selling investments and securities during the stock market's boom years of the 1920s. In 1927, he was ready to try again—but this time he added an improvement to his original invention. The new model sliced the bread and wrapped it in plastic to keep it fresh. By 1933, bakers across the country reported that most of their customers were asking for sliced bread. That same year, Rohwedder sold his invention to the Micro-Westco Company of Iowa and took a job as vice president and sales manager of the company's Rohwedder Bakery Machine Division.

An early bread-slicing machine

PENICILLIN

Alexander Fleming (1881–1955), a Scottish physician, had been studying and lecturing on bacteriology when his work was interrupted by World War I. During his service as a captain in the Army Medical Corps, Fleming saw firsthand how many soldiers suffered and died from infected wounds. After the war, he returned to St. Mary's Hospital in London when he renewed his interest in bacteriology.

In 1928, Fleming went off on a two-week vacation but didn't bother to clean up his laboratory before he left. Once he came back, he saw one especially disgusting specimen on a glass plate, on which he had left a staphylococcus culture. A greenish yellow mold had sprouted on the culture, which was nothing remarkable considering it had been sitting in the open air for two weeks—but what was strange was a bacteria-free zone that encircled the staphylococcus. He ran tests on the mold and found that it had been created, not by a staphylococcus culture that had gone bad, but by Penicillium notatum. No one in Fleming's lab was studying Penicillium notatum but researchers were in the lab one floor below. A spore of Penicillium notatum must have been carried into his lab and found its way to the staphylococcus culture. Vastly more important than the arrival of the wayward spore was what it did: This Penicillium notatum stopped the growth of bacteria on part of the glass plate.

Fleming named the substance penicillin and published his discovery in the *Journal of Experimental Pathology*; incredibly, the medical community overlooked it for nearly ten years. Only in 1939 was penicillin resurrected by an Australian physiologist, Howard Florey, who was leading a research team at Oxford University to find medicines that would kill bacteria. After giving laboratory mice a lethal dose of bacteria, Florey and his colleagues injected the mice with penicillin. The results were astounding: In almost every case, the mice were cured.

Immediately, the British and American medical communities began to manufacture penicillin. It was used to treat pneumonia, diphtheria, and scarlet fever, among other infectious diseases, and during World War II it saved the lives of millions of soldiers and civilians. Penicillin is the most widely used antibiotic worldwide and manufacturers have also introduced semi-synthetic versions.

Alexander Fleming, the discoverer of penicillin, at work in his laboratory

Chain Saw

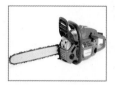

Why does a chain saw slice through tree trunks, thick brush, and other obstacles so easily? The answer is horsepower and the miracle of the two-stroke engine. A standard-size chain saw is smallish, with a blade or cutting bar 16 inches long, and a weight of about 4 pounds. Yet the engine of this little machine revs up to about 3 horsepower because the engine is air-cooled and two-stroke rather than four-stroke. An air-cooled engine does not need a radiator, water pump, or water to cool it, which cuts out a significant amount of weight in a hand-held gasoline-powered tool. As for the chain saw's two-stroke engine, it produces power at every rotation of its crankshaft, while a four-stroke fires up on every other rotation of its crankshaft—this is what accounts for the extra boost of power in a two-stroke engine, and why it is used in jet skis, dirt bikes, leaf blowers, and small outboard motors.

The man who built and patented the first gasoline-powered chain saw was Andreas Stihl (1896–1973), a Swiss engineer. After receiving his patent for his "tree-felling machine" in 1929, Stihl started a company to manufacture chain saws. By 1931, he was exporting Stihl chain saws to the United States and Russia. They are so light, easy to use, and fast, they have virtually displaced all hand-powered saws in the lumber and forestry businesses.

On a traditional saw, the teeth are cut into the blade and they do not cut a thing until human muscle forces them into the wood and then back again, repeating the motion until the saw has cut through the tree trunk or limb. It's the chain of the chain saw that makes all the difference. The chain runs along a groove on the blade or cutting bar; the saw's teeth are on the chain. It is the speedy rotation of the sharp chain that makes it so easy to cut through wood—the engine moves the teeth, rather than your elbow grease.

A logger uses an early chainsaw

Sunglasses

It is said that in fourteenth-century China, judges wore dark-colored glasses so no one in the courtroom could observe their reaction to any testimony. In Europe, from the fifteenth through the eighteenth centuries, inventors had produced smoked and tinted eyeglasses which they believed would correct vision impairments. It never occurred to them that such glasses could shield the eyes from the sun's glare (which was the only thing the tinted lenses were good for, as colored glass did nothing to improve anyone's vision).

On a bright sunny summer day in 1929, Sam Foster sold the first sunglasses from his counter in the Woolworth store on the boardwalk in Atlantic City, New Jersey. They were an instant hit with the sunbathing public. Grant's timing was excellent because, that same year, Edwin Land invented a cellophane filter to polarize light and reduce glare. In 1932, Land and a Harvard physics professor, George Wheelwright, teamed up to manufacture sunglasses coated with what they called the Polaroid filter.

Soon the U.S. military became interested in Polaroid sunglasses as a way to protect pilots from the bright glare they often experienced when flying at high altitudes. Researchers at Bausch & Lomb discovered that lenses tinted a dark green reduced glare significantly. In 1936, Ray–Ban (founded by B&L) introduced a new style of sunglasses for aviators with larger, curved lenses that dropped below the pilot's eyes; this new style shielded the pilot from glare even when he glanced downward to read his instruments panel. This new style of sunglasses became known as Ray-Ban Aviator glasses.

Sunglasses, whether aviators or standard models, made the wearer look cool. As actors, musicians, and other celebrities were photographed wearing sunglasses, the market for them exploded. James Dean popularized his trademark Ray-Ban Wayfarers; John Lennon is forever associated with Teashades, little round lenses set in thin wire frames; and very large sunglasses are known as Jackie O's, a tribute to Jacqueline Kennedy Onassis, who wore them in the late 1960s and early 1970s.

Oversized sunglasses were a Jackie Onassis trademark

Saran Wrap®

Ralph Wiley was a college student who worked part-time cleaning glassware at the Dow Chemical laboratory. One day, he came across a glass vial that contained some type of hardened residue that no amount of soap or scrubbing could dislodge. He showed it to some of Dow's chemists and eventually they were able to replicate the substance as a foul-smelling, oily, green anti-corrosive liquid. They called the stuff "Saran," and found an instant market for it with the U.S. military—they applied Saran on aircraft to protect the wings and fuselage from the caustic effects of sea spray.

Saran was an ideal preservative thanks to its tightly bound molecular structure, which created an almost perfect seal against moisture, oxygen, and a host of chemicals.

After World War II, Dow chemists managed to make improvements in Saran which eliminated the sickly green color and the nasty odor. By 1953, Saran was approved for packaging food and Dow introduced Saran Wrap® as the first plastic wrap designed to make a virtually airtight seal over or around food. Before Saran Wrap, leftovers were wrapped in wax paper, which did not keep foods fresh for very long, did not seal in or keep out odors, and did not cling to the bowl or dish in which the food was stored. In 1947, the Reynolds Metals of America company introduced sheets of aluminum foil as thin as 1/7,000th of an inch. It was more durable than wax paper and crinkling the edges of the foil made a much better seal than waxed paper. However, once Saran Wrap appeared on the market, aluminum foil dropped back into second place. It remains there to this day.

Saran Wrap revolutionized food packaging

Answering Machine

A Danish inventor and telephone engineer, Valdemar Poulsen (1869–1942), is often credited with inventing the first answering machine. He magnetized a piano wire, which he used to record telephone conversations but since Poulsen's device did not answer calls, it was not a real answering machine.

The first automatic machine that took phone calls was invented in 1935 by Willy Müller of Switzerland. It stood 3 feet tall and found a market among Orthodox Jews who are forbidden to answer the telephone on the Sabbath.

In 1954, Dr. Kazuo Hashimoto (1921–1995) of Japan began working on the development of a compact answering machine for homes and businesses. His ANSA FONE machine was the size of a tape recorder which made it much more practical than Müller's model. It went on sale in the United States in 1960. ANSA FONE is just one of more than 1000 patents—800 of which were telephone-related—that Dr. Hashimoto was awarded in his long and productive career. His ANSA FONE is part of the collection of the Smithsonian Institution.

Building upon Dr. Hashimoto's work, the next generation of answering machines became more compact and could store more messages. PhoneMate, which debuted in 1971, weighed only 10 pounds and could hold twenty messages on reel-to-reel tape; it included an earplug so you could listen to your messages privately. In the early 1980s, Dr. Hashimoto introduced the first digital answering machine and invented the popular feature caller identification, or caller I.D.

Gordon Matthews (1937–2002) of Texas invented voicemail in 1979. By the 1970s, more business was being conducted via telephone than ever before but corporate efficiency experts estimated that only 1 in 4 calls made to a business office actually reached a human being. Installing a separate answering machine for every telephone in a large company would be prohibitively expensive; what was needed was centralized answering system—in other words, voicemail. To market his invention Gordon Matthews founded VMX—Voice Message Express—and sold his first voicemail system to 3M.

A modern answering machine

Nylon

Throughout the 1930s, tensions between the United States and the fascist government in Japan were having an impact on trade. As the Japanese began to restrict their export of silk to the United States, ladies' silk stockings became increasingly harder to find. The DuPont chemical company already had a team in place working on synthetic fibers. The team's chief researcher, Wallace Carothers (1896–1937), had been instrumental in developing a synthetic rubber called neoprene. Now Carothers and his fellow researchers began to look for a synthetic fiber to replace silk.

Natural rubber and natural silk are strong and resilient because they are composed of long strings of molecules known as polymers. Carothers and his colleagues had already developed a synthetic polymer that made neoprene, now they worked to find a synthetic polymer that would act like silk. At the time, the standard method of testing synthetic polymers at DuPont was to run them through a device popularly known as "the spinner" to see if the polymer could be spun into thread. Carothers's synthetic polymer ran into trouble because heat was involved in the spinning process and when his polymer came in contact with heat, it melted.

One day, a member of the research team name Julian Hill stuck a glass rod into a beaker of another type of polymer known as polyester. The polyester clung to the rod and when Hill drew it from the beaker it came out in long threads. This method is known as cold-drawing, distinct from the heat involved in the spinner. It's said that one night, after Carothers and the other department chiefs had gone home, Hill and some friends took a beaker of polyester out to a hallway. They did a standard cold-drawing and then gently began to stretch the thread of polyester to see how far it would go. To their astonishment and delight, it stretched the entire length of the corridor. They went back to the synthetic silk polymer, performed a cold-drawing of it, and found it was as flexible as polyester and it was stronger than natural silk.

This new synthetic fiber became nylon, which revolutionized the stocking industry and even became a means of exchange during World War II, when soldiers traded nylon stockings for a wide variety of goods and services.

Wallace Hume Carothers, the inventor of nylon, in the DuPont laboratory

Parking Meter

Park-o-meter, 1935

In 1935, the traffic committee of the Oklahoma City Chamber of Commerce tackled an especially vexing problem: employees of various offices and businesses near the downtown area were taking all the parking spots in the city center and leaving their cars there all day, to the great inconvenience of people who had driven downtown to shop. Traffic committee member Carl C. Magee came up with a solution: charge by the hour for every downtown parking space, and place a small machine at each parking space to collect the money. Office workers would not be able to run down to their cars every hour "to feed the meter," and once they parked elsewhere the parking spaces would be free for shoppers and other visitors. And there was the additional benefit of the meters generating income for the city. The city officials liked Magee's idea and installed the first parking meter on July 16, 1935 (Magee did not apply for a patent until 1938).

Magee's meter was a small box mounted on a metal pole. The driver inserted a coin in a slot and turned a knob which displayed how long the car could "rent" that space. As a solution to Oklahoma City's downtown parking problem and an easy way to collect revenue, the parking meters were an instant success at city hall. Magee started a company to manufacture parking meters for other cities and towns. Not everyone who encountered parking meters was wildly enthusiastic—citizens of some towns in Texas and Alabama resented being charged for the privilege of leaving their cars on their own city streets and destroyed the meters. (In the movie classic, *Cool Hand Luke*, Paul Newman's character is sent to prison for decapitating parking meters.)

Parking meters also generate a second source of income for cities—parking tickets issued to drivers who have overstayed their allotted time. In 1960, New York City hired an all-female force of parking attendants to ticket vehicles whose meters had run out. Known as "meter maids," the group refused to hire men until 1967.

A line of early parking meters in Omaha, Nebraska, 1938

Radar

In 1934, the chairman of Britain's Committee for the Scientific Survey of Air Defense called on Robert Watson-Watt, chief of the National Physical Laboratory. There was a rumor that Nazi Germany had developed a "death ray" using radio waves that could blow up enemy aircraft. Could Watson-Watt make a death ray, too? It sounds like the plot of science fiction novel but Watson-Watt did make some calculations and was happy to report back to the air defense committee that it was impossible for radio waves to generate enough power to shoot planes out of the sky. Members of the committee may have been crestfallen that Britain would not have a death ray but, on the other hand, it meant the Nazis did not possess a radio wave death ray, either. There was more good news—Watson-Watt had found that radio waves could be used to detect any enemy aircraft en route.

Radio waves travel very fast, about 1,000 feet per microsecond. When a radio wave bounces off an object, it gives off an echo. Based on how long it took for the echo to be picked up, the radar transmitter can calculate how far away the object is, in what direction it is moving, and how fast it is traveling. An early demonstration of radar showed that radio waves could detect the presence of a plane that was 40 miles away. In other words, radar would give the British advance warning of a Nazi air raid. By the time World War II broke out in 1939, the British government had built twenty radar stations along the coast, from the Isle of Wight to Dundee in Scotland.

The British called this innovation "radio direction finding" or RDF. It was the Americans who gave Watson-Watt's discovery the name that would stick—radar, shorthand for "radio detection and ranging."

Radar proved to be vital during the Battle for Britain, when Nazi planes bombed British cities and towns almost nightly. Not only did it give warning to civilians to take cover in air-raid shelters, it also alerted the Royal Air Force where their planes would be needed most. In 1942, King George VI knighted Watson-Watt.

Radar inventor Robert Watson-Watt performs an experiment with a kite and wireless transmitter

SUNSCREEN

For centuries, a suntan was a kind of social stigma, a sign that you were a member of the laboring class who worked outdoors all day. A pale, untanned complexion was the sign of a lady or a gentleman. Then, in 1920, Coco Chanel returned home tanned after a cruise on a friend's yacht. As the most influential woman in the fashion industry, if Chanel had a tan then *everyone* must have a tan. Hollywood stars, both women and men, embraced the bronzed look and tanning became chic.

Of course, while some people tan under the sun, other people burn, and to assist the latter group, a French chemist and founder of the L'Oréal line of cosmetics, Eugene Schueller (1881–1957), invented the first sunscreen in 1936.

The sun emits three types of ultraviolet rays: UV-A rays will give you a tan but will also damage your skin cells; UV-B rays are the ones that give you sunburn; and UV-C rays are filtered almost completely by the Earth's atmosphere and do no harm to the skin. When shopping for sunscreen, look for products that will protect you against UV-A and UV-B rays.

Sunscreen consumers should read the labels for the SPF, which stands for sun protection factor. The SPF number indicates how long you can stay out in the sun after applying the sunscreen. For example, if the label reads "SPF 15" and you start to get a sunburn in ten minutes, then multiply ten minutes by fifteen and you'll know how long you'll be protected. The higher the SPF, the greater the protection factor. That's the theory, anyway. Dermatologists suggest that if you plan to be in the sun for a prolonged period, then reapply sunscreen about every two hours.

Sunscreen reflects the sun's ultraviolet rays. Sunblock, however, is a different product and does what its name suggests—it blocks ultraviolet rays from reaching the skin.

Coco Chanel, the arbiter of taste in the 1930's, made tanning a necessity for fashionable people

ALUMINUM SIDING

Aluminum being produced in sheets

In the 1920s and 1930s, the Sears, Roebuck catalog offered sheets of steel siding (stamped to look like brick or stone) for buildings but there was no demand for it—the weight of the siding may have had something to do with its inability to find a wide market. Aluminum, the shiny, lightweight new metal of the twentieth century, found an enthusiastic market among architects who created elaborate designs for aluminum panels, which they mounted on New York City's Chrysler Building and the Cathedral of Learning. These uses of aluminum were monumental in scale and, at the time, it appeared that no one imagined there might be more mundane uses for the metal.

In 1937, machinist Frank Hoess of Indiana began experimenting with metal siding for small buildings such as homes and shops. He wanted to use a metal that would not warp, rust, or split over time. Hoess began with steel siding but eventually chose aluminum as the best solution. He also produced his siding to look like traditional wooden clapboards. During World War II, the armed forces virtually took over all metal plants to produce all the metal equipment necessary to defeat Germany and Japan. After the war, however, Hoess started a company to sell his aluminum siding. The potential was huge: A postwar building boom in the suburbs required construction companies to look for new, cheaper, more efficient methods to build houses. Along with wallboard for interior walls and asbestos insulation, aluminum siding joined the top three new building materials.

Unfortunately for Hoess, he could not compete with such giants as Reynolds Aluminum, which soon dominated the siding market. Reynolds's advertising campaigns focused on the convenience of aluminum siding (it never required painting) as well as its homey look (it was sold just like clapboard in 8- or 12-foot lengths and it provided a lifetime of "traditional colonial beauty").

NONREFLECTING GLASS

Katharine Burr Blodgett's (1898–1979) aptitude in science won her a scholarship to Bryn Mawr College where she studied mathematics and physics. After graduating from Bryn Mawr, she went on to the University of Chicago for a master's degree in physics—which she completed in only one year. Just twenty years old and fresh from Chicago, she was hired by General Electric—the first woman research scientist to work in the company's laboratory.

At the GE lab, Blodgett worked with Irving Langmuir, a future Nobel Prize winner. Together they generated a flurry of patents for improvements to everything from vacuum pumps to light bulbs. Langmuir had found that oily substances left a film on the surface of water, one-molecule thick; he wondered if there was a practical application for such a film, and asked Blodgett to see what she could make of it. Blodgett discovered that repeated applications of an oily liquid soap to a sheet of glass would eliminate glare. Specifically, it took forty-four coats, each one molecule thick, of the oily soap to block glare. Interestingly, although the treatment eliminated glare completely, 99 percent of light still passed through the glass. Blodgett had invented nonreflecting glass. The problem with Blodgett's invention, however, was that the coating was easy to rub off. It would be other researchers who, working with Blodgett's initial results, found a way to apply a more durable coating on glass.

During World War II, Blodgett invented a device that gave off an artificial smokescreen on the battlefield and she devised better ways to de-ice the wings of fighter planes.

She spent almost all of her life in Schenectady, New York, the place where she had been born. Before her death at age eighty-one, her hometown celebrated their most famous citizen with Katharine Blodgett Day.

Nobel Prize–winner Dr. Irving Langmuir with radio inventor Guglielmo Marconi

PHOTOCOPIER

Rub a balloon on a wool sweater and you create static electricity, which will perform little "tricks," such as attracting bits of paper to itself. A photocopier works in the same way. Inside the machine sits a drum charged with electricity and a cartridge filled with fine black and/or colored powder known as toner. The charged drum will attract the toner powder the same way the static electricity of a balloon will attract tiny pieces of paper.

There is also an incandescent or fluorescent bulb inside the copier (UV bulbs are too powerful and could damage the eyes). The light strikes the sheet of paper that's placed on the copier's surface and the text or images are reflected onto the drum. The drum attracts the bits of toner and when a sheet of copy paper passes over the drum, it picks up the toner. The heat from the light bulb fuses the toner to the page as the copy starts to slide out of the machine and into the tray.

The man who invented the photocopier was Chester F. Carlson (1906–1968), a physicist who had also studied electronics. Fresh out of college, he had taken a job in the patent department of a battery manufacturer, P. R. Mallory and Company. A chronic headache for those who worked in the patents department was the task of keeping enough copies on hand of patent drawings and applications. While others thought photography was the solution, Carlson began to experiment with static electricity. He produced his first instant copy in 1938 and began shopping his invention around. Twenty different companies turned him away. Finally, in 1947, the Haloid Company (known today as Xerox) made Carlson an offer for the commercial rights to his invention. Xerox began marketing its first photocopier eleven years later, in 1958.

1938

A New York detective in 1950 uses the "identiscope" which creates photostatic copies of documents.

Shopping Cart

Sylvan Goldman (1898–1984) once observed that food is something everybody needs, but they can use it only once. For that reason he spent this whole career in the food and supermarket business. He started out as a dealer in wholesale produce. Then he opened a string of self-service grocery stores in his home state, Oklahoma. He sold his chain to Safeway but lost all of his Safeway stock in the crash of 1929. He found a management job with the Piggly Wiggly stores and by the mid 1930s was part owner of the chain.

In his self-serve grocery stores, Goldman had supplied his shoppers with shopping baskets. They were convenient but limited—there was only so much stuff that could be crammed into a hand basket. How could he get his customers to buy more products on each visit to the Piggly Wiggly? One night in his office he glanced at a wooden folding chair and he had an inspiration: why not put a handle on top, wheels on the legs, and a basket where the seat was? Better yet, build a cart with two baskets—one on top of the other.

He took his idea to a mechanic named Fred Young who built the prototype; a simple metal frame mounted on four wheels with the two wire baskets. Since space was also a consideration, Young designed the cart so the baskets tilted up and nested together. It was simple and wonderful but the customers wouldn't use the shopping carts. Women said they were unattractive. Elderly people thought the carts made them look feeble. And as for the men, well, only a guy who was too weak to carry his groceries pushed them around in a cart. To overcome his shopper's prejudices, Goldman hired models, men and women, of various ages, to push the carts up and down the aisles of the Piggly Wiggly pretending to shop. And he installed a greeter at the supermarket door to welcome shoppers and encourage them to use a cart. The ruse worked.

America's first national supermarket, complete with shopping carts

TEFLON®

In spring 1938, at the DuPont laboratory in Jackson, New Jersey, Dr. Roy J. Plunkett (1911–1994) was working on developing a nontoxic gas that could be used in refrigerators. At the end of the working day, he mixed up in a cylinder of what he thought was Freon gas but the next morning, when his assistant, Jack Rebok, opened the cylinder's valve no gas came out. From the weight of the cylinder, Plunkett and Rebok knew something was in there, so they cut it open. Settled at the bottom was an unknown waxy white mass.

As Plunkett and his colleagues began to test the substance, they learned that it was a polymer, like the substances DuPont was using to create synthetic fabrics. They also discovered that it was inert, meaning that it was impervious to heat, electricity, and even acid. The mysterious substance had one other quality—nothing stuck to it.

The first two uses for Teflon® could not have been more extreme: It was used in the first atomic bomb to protect gaskets from being corroded by uranium—and it was used to coat bakery muffin pans.

The first Teflon-coated nonstick frying pans were not put on the market until 1960. They were a marvel but one had to be careful with their Teflon cookware because it scratched easily, and once it was scratched that portion of the pan was no longer stick-free.

Thanks to its miraculous nonstick surface, Teflon has entered the English language in another sense, too. Ronald Reagan and Bill Clinton became known as "Teflon presidents" because no scandal stuck to them. Mobster John Gotti was known as "the Teflon Don" because he dodged a conviction on racketeering and assault charges (thanks to his informants in the police department, who kept him one step ahead of investigators).

As for Roy Plunkett, since 1988 DuPont has bestowed The Plunkett Award on individuals who have found innovative new uses for Teflon.

Teflon was used to coat gaskets in the first atomic bomb

TOOTHBRUSH

Early toothpaste tubes were made of metal, and could be exchanged at drugstores

At least as early as 3000 B.C., humans cleaned their teeth by rubbing them with the frayed end of a twig. The Romans even had a kind of toothpaste, although it was so abrasive it damaged tooth enamel. Around the year 1000 A.D., the Chinese mounted bristles from the back of a hog's neck on a small brush made of bamboo or bone and used it to clean their teeth. Toothbrushes with animal bristles remained the norm into the early years of the twentieth century. Then, in 1938, DuPont began to market a toothbrush with bristles made of nylon, the new synthetic fiber that had been invented in the company's lab.

A variety of toothpastes have been used over the centuries, including pastes that contained such ingredients as emery powder, soap, and chalk. In 1873, Colgate introduced toothpaste in a jar. In 1892, Dr. Washington Sheffield of Connecticut introduced toothpaste stuffed into a collapsible tube. He marketed it as "Dr. Sheffield's Creme Dentifrice."

The market for toothbrushes and toothpaste was small in the United States—before World War II, very few Americans brushed their teeth. During the war, however, U.S. troops were given toothbrushes and toothpaste and ordered to clean their teeth as part of their daily hygiene regimen. When the troops returned home, they introduced the habit to their families. Today, dentists recommend that we brush and floss twice daily.

The first electric toothbrush came on the market in America in 1960; it was produced by Squab and was called the Broxodent. A year later, General Electric manufactured the first cordless, rechargeable battery operated toothbrush. And in 1987, Interplak brought out the first electric toothbrush with rotary action. Yet despite these innovations, dentists have found that the electric brush is no better at cleaning teeth than the old-fashioned manual variety.

Bob Hope shows soldiers in 1942 proper toothbrushing technique using his finger

BALLPOINT PEN

László Biró (1899–1985) couldn't decide what he wanted to be when he grew up. He entered medical school but didn't graduate. He practiced hypnotism for a time but gave it up for a regular salary as a clerk at an oil company. In search of thrills, he drove racecars. Finally, he took a job as a journalist. Now that he was writing all day long, Biró discovered firsthand the frustrations of working with a fountain pen—the ink took so long to dry that one careless move of the hand could smudge a whole paragraph. The ink used in the newspaper's printing plant dried on contact but newspaper ink wouldn't work in a fountain pen; it was too thick. So Biró invented a new kind of pen that would work with quick-drying ink.

Biró's pen contained a cylinder of viscous, quick-drying ink. At the tip was a tiny steel ball. As the point of the pen moved across a sheet of paper, the tiny ball rotated, collecting ink from the cartridge and depositing it on the paper where it dried virtually on contact. By 1939, Biró had perfected his new pen. That same year, Hitler began his conquest of Europe and Nazi sympathizers in Hungary passed a series of anti-Jewish legislation. Biró, who was Jewish, fled with his family to Paris. When the war caught up with them in France, the Birós emigrated to Argentina. There, he patented his ballpoint pen.

To Biró's surprise his pen became his contribution to the war effort. The United States Department of Defense had been searching for a pen that could be used in aircraft. The thin ink in fountain pens acted erratically at high altitudes and, of course, there was still the smudging problem. The solution was Biró's ballpoint. The thick ink was impervious to altitude and it dried almost on contact. Throughout the war, American pilots rarely used any pen but ballpoints.

László Biró remained in Argentina for the rest of his life. Every year, the country celebrates Inventors' Day on Biró's birthday—September 29.

A B-47 jet pilot uses a ballpoint pen during flight in 1954

HELICOPTER

As a young man growing up in Kiev, Ukraine, Igor Ivanovich Sikorsky (1889–1972) had followed with enthusiasm the successful flights of the Wright brothers and Count Zeppelin. He studied engineering as it related to the infant science of aviation in Paris, and then returned home to Kiev to build aircraft. It was Sikorsky who helped the world to see flight as a new mode of travel, through his designs for luxury passenger airplanes. In the last days of czarist Russia, he designed "The Grand," the first deluxe multiengine commercial aircraft. When World War I broke out, followed by the Bolshevik Revolution in his homeland, Sikorsky traveled to France where he worked designing bombers.

After the war, Sikorsky emigrated to the United States, opened his own passenger aircraft manufacturing plant on Long Island, and began supplying airplanes to Pan Am for its runs to destinations in the Caribbean and South America. Sikorsky's company manufactured the "American Clippers," which flew across the Atlantic and the Pacific.

While he designed elegant passenger aircraft, Sikorsky worked in private on designs that would become the first successful helicopter. In 1939, his designs were complete and he constructed his first helicopter, piloting it himself. The problem with earlier models of helicopters had been to find some way to counteract the powerful torque of the top rotor of the helicopter. Sikorsky solved the problem by mounting a single rotor on the tail—it delivered enough thrust to balance the action of the top rotor. Weeks after Sikorsky made a test flight with helicopter, Hitler invaded Poland and World War II began.

By the time the war ended in 1945, the U.S. military had purchased 400 of Sikorsky's helicopters, using them especially for life-saving missions in settings where no airplane could land.

For his achievement, Sikorsky was awarded the National Medal of Science and the Wright Brothers Memorial Trophy and he was inducted into the International Aerospace Hall of Fame. His Sikorsky Aircraft Corporation still builds helicopters in Stratford, Connecticut.

Igor Sikorsky, inventor of the helicopter, demonstrates his flying skill as a helper puts a briefcase into the basket mounted on the helicopter's nose

Paperback Book

James Hilton's *Lost Horizon* was one of the first paperbacks ever published

The Egyptians had papyrus scrolls 5,000 years ago. By 590 A.D., Irish monks bound manuscripts between soft vellum covers. And Ann Sophia Stephens published the first paper-covered dime novel in 1859. But the modern paperback book was born in 1939 when publisher Robert F. de Graff founded Pocket Books, the first paperback-only publishing company in the United States. De Graff believed that the American reading public would buy paperbacks if they were economical, readily available, and reasonably sturdy.

He started with ten titles: *Lost Horizon*, by James Hilton; *Wake Up and Live*, by Dorothea Brande; *Five Great Tragedies*, by William Shakespeare; *Topper*, by Thorne Smith; *The Murder of Roger Ackroyd*, by Agatha Christie; *Enough Rope*, by Dorothy Parker; *Wuthering Heights*, by Emily Brontë; *The Way of All Flesh*, by Samuel Butler; *The Bridge of San Luis Rey*, by Thornton Wilder; and *Bambi*, by Felix Salten. The books sold for 25 cents each, and could be found in drugstores, department stores, and newsstands, as well as bookstores.

De Graff's hunch was correct and soon he had competition. In 1941, Avon Books began publishing just paperbacks. In 1945, Bantam published its first paperback—a paperback edition of F. Scott Fitzgerald's *The Great Gatsby*. But it was not until 1976 that *Publishers Weekly* acknowledged the trend and compiled its first list of paperback bestsellers. The top five on that initial list were *Helter Skelter*, by Vincent Bugliosi with Curt Gentry; *The Furies*, by John Jakes; *Centennial*, by James Michener; *The Ultra Secret*, by F. W. Winterbotham; and *Aspen*, by Burt Hirschfield.

In 1974, paperbacks got their own Book-of-the-Month-style club with the founding of the Quality Paperback Book Club.

Today, paperback books are ubiquitous and readers are quick to snap up softcover editions of everything from Shakespeare's plays to massive *Harry Potter* paperbacks.

Paperbacks are cheaper and easier to carry than hardcover books

DUCT TAPE

People who call this product "Duck" Tape aren't guilty of mispronunciation—they are in fact calling the tape by its original name. In 1942, the U.S. military developed this heavy-duty cloth-covered tape to make ammunition cases watertight. Since water rolled off it (the way water rolls off a duck's back), the troops called it "Duck Tape."

It's original color was different, too—drab Army green. But one thing about the tape has remained consistent—it is extremely versatile. Soldiers used it for everything from covering holes in their boots to performing quick emergency repairs on Jeeps and even aircraft.

After the war, during the housing construction boom in the suburbs, contractors used Duck Tape to make a waterproof seal over heating and air conditioning ductwork. Since the ducts were silver, the Johnson and Johnson Permacel Division, which had invented the tape, changed its color to gray. At that point, since the tape was being used primarily on ductwork, it became Duct Tape.

What makes Duct Tape so strong? It is made up of three layers. On the bottom is the adhesive. The next layer up is fabric mesh. The top layer is durable, waterproof plastic (Polyethelyne). It is incredibly strong tape. In one experiment, a length of doubled-over Duct Tape was used to haul a 2,000-pound automobile out of a ditch. Yet you can tear Duct Tape easily with your hands.

In recent years, the evangelists for Duct Tape and its myriad uses have been Jim Berg and Tim Nyberg, coauthors of books including *The Ultimate Duct Tape Book*, 365 Days of Duct Tape Page-A-Day calendars, and *Duct Shui*, Berg and Nyberg's meditation on the philosophical aspects of Duct Tape.

1942

Shop owners used duct tape to strengthen their glass windows during WWII

AQUALUNG

A sixteenth-century diagram of an underwater breathing apparatus

Marine exploration had always been hampered by one thing—human beings cannot breath underwater. As far back as Aristotle, various inventors had suggested numerous contraptions to enable swimmers to breath underwater but none of them were reliable. It was not until 1943 that a practical solution was developed by two Frenchmen, Jacques-Yves Cousteau (1910–1997) and Emile Gagnan (1900–1979). They called their invention the Aqualung. The genius part of their invention was the attachment of a regulator to the portable oxygen tank: This regulator adjusted air pressure automatically so that the air pressure in a diver's lungs was equal to the pressure of the water. Cousteau and Gagnan's air pressure regulator gave the world what it never had before—a safe, reliable device that enabled divers to remain underwater for prolonged periods of time.

In 1943, France was still under Nazi occupation, so Cousteau and Gagnan did not bring their Aqualung on the market until 1946, a year after the end of World War II. By 1951, the Aqualung was for sale in Great Britain and Canada and, the following year, a company called U.S. Divers acquired distribution rights for the Aqualung in the United States.

Gagnan emigrated to Canada where he continued inventing new devices for scuba equipment. Putting their invention to excellent use, Cousteau became a world-renowned underwater explorer and marine conservationist. But the Aqualung also gave birth to a new recreational activity—scuba diving. Diving schools can be found around the globe and it is quite common for seaside resort hotels to offer their guests scuba lessons and escorted underwater tours led by professional divers.

1943

Aqualung inventor and oceanographer Jacques Cousteau

Atomic Bomb

The most fearsome weapon the world has ever known evolved from a succession of innocent scientific discoveries. While experimenting with uranium in 1896, Frenchman Henri Becquerel (1852–1908) discovered that when an unstable atomic nucleus loses energy, it emits particles or electromagnetic waves know as radiation. In 1899, New Zealander Ernest Rutherford (1871–1937) found that two types of radiation are emitted by uranium, alpha particles and beta particles, and that the alpha particles in particular released a very concentrated level of energy.

A German, Walther Bothe (1891–1957), took the first step toward drawing energy from the nucleus of an atom. Meanwhile, similar work was being done by a Hungarian physicist who had settled in the United States, Leo Szilard (1898–1964). By the late 1930s, physicists believed that nuclear fission—a process that used a neutron to split the nucleus of an atom in two—would release an enormous, even destructive, amount of nuclear energy.

In August 1939, less than a month before the Nazi invasion of Poland and the start of World War II, Szilard sent a letter (co-signed by Albert Einstein) to President Franklin D. Roosevelt warning him that the Nazis were attempting to create an atomic weapon. Determined to manufacture an atomic bomb before the Nazis, FDR authorized what became known as the Manhattan Project, an immense scientific-governmental-military operation that involved 37 installations in 19 states and Canada, employed 43,000 employees, and had a budget of $2.2 billion. By summer 1945, the project had developed three atomic bombs; the first was tested near Alamogordo, New Mexico, on July 16, 1945. After that successful test, the other two were used to destroy the Japanese cities of Hiroshima on August 6 and Nagasaki on August 9, 1945.

The bomb the physicists of the Manhattan Project developed contained a sphere of uranium-235 and a core of plutonium-239 surrounded by high explosives. The bomb was detonated in four steps: the explosives were fired; the shock waves from the explosion compressed the plutonium core; the compression of the plutonium began the process of nuclear fission; and then the atomic bomb exploded. It all happened very quickly—fission occurred in about 560/billionths of a second.

Albert Einstein and Leo Szilard with the letter they sent to Franklin D. Roosevelt warning him of Nazi Germany's plans to build an atomic bomb

MICROWAVE OVEN

A magnetron, the primary piece of machinery in a microwave

The first microwave oven stood 5 feet, 6 inches tall and weighed 750 pounds. A countertop model was not available. In fact, the first microwave ovens were sold not to families but to restaurants, ocean liners, and railroad dining cars, where large amounts of food had to be cooked or reheated quickly.

The microwave's inventor, Percy L. Spencer (1894–1970), was an orphan from Maine who had been forced to quit school at age twelve and take a job in a mill. When he joined the U.S. Navy, Spencer was able to jump-start his education. He went to work for Raytheon, and during World War II was instrumental in producing combat radar so sensitive it could detect the periscope of a Nazi U-boat.

One day, before the end of the war, Spencer was working with a magnetron for a radar set, when he noticed that the chocolate bar in his pocket was starting to melt. He suspected that microwave radiation from the magnetron was giving off enough heat to melt the chocolate. He wondered if it could also cook, so he placed kernels of popping corn in front of the magnetron—popcorn popped all over the room.

Raytheon received a patent for Spencer's microwave cooking method in 1945, and introduced their first microwave oven to the market in 1947; they called it the Radarange; it required 3,000 watts of power. The microwave suffered from other setbacks: It did not brown meat and it could not produce crispy French fries—but Raytheon's chairman, Charles Adams, believed in the product. When Adams demanded that his personal cook use only a microwave oven to prepare meals, the cook quit.

To reach a wider market, Raytheon developed a 220-volt wall unit microwave but the retail price of $1,295 killed sales. It was not until 1967 that Raytheon was able to produce a more economical counter-top microwave oven—and even that one cost $495.

A chef prepares food with the first microwave, the Raytheon Radarange

KITTY LITTER

Edward Lowe's (1920–1995) father manufactured industrial absorbents. After serving in the U.S. Navy during World War II, Lowe returned home to Minnesota and joined his father's company as a salesman. At the time, most cat owners put out boxes of wood ash for their pets, but Lowe's neighbor, Kay Draper, complained that after using the box her cats tracked ash all over the house. Lowe suggested that she use one of the products manufactured by his family's company, absorbent clay pellets known as Fuller's Earth. Draper tried it and was delighted with the results. The pellets absorbed the smell and the moisture of cat urine and did not cling to the cat's paws when it climbed out of the box.

Inspired by Draper's success, Lowe began packaging the clay pellets in 5-pound bags and selling them to stores as "kitty litter." The retail price for a 5-pound bag was 65 cents. To build a market, he filled the trunk of his car with bags of kitty litter and drove around the country, personally selling the product to store owners. To get attention from cat owners, he attended a cat show where he cleaned cat boxes and replaced the messy wood ash with his tidy kitty litter. He founded his own company, Edward Lowe Industries, to make and sell his kitty litter, and in 1964 introduced Tidy Cat® to the American market. By 1990, Lowe's company was selling $210 million worth of kitty litter annually.

1946

Edward Lowe's target audience for kitty litter

TRANSISTOR

In 1947, Alexander Graham Bell's patents for the telephone were starting to expire, prompting a situation that raised the anxiety level among the executives at American Telephone and Telegraph (AT&T). To bail out the company, they brought their former president and proven problem-solver, Theodore Vail, out of retirement, begging him to find a solution that would keep the company solvent. Vail's solution was to make AT&T the first telephone company in America to provide transcontinental phone service. To make good on its promise, AT&T installed amplifier tubes (which augmented the signal as it passed along the line) in every switch box from one end of the country to the other. It worked but it wasn't ideal. The amplifiers burned out quickly and used up too much power but there was no alternative—until just before Christmas 1947.

Beginning in 1945, the research division of AT&T, Bell Labs, had assembled a team of physicists, chemists, and engineers to find a more efficient method for sending telephone signals across the country. John Bardeen (1908–1991) and Walter Brattain (1902-1987) knew they needed two contact points, one that would receive an incoming signal, the other that would strengthen the current as it passed it along the line. Two years of experimentation had taught them that they would need two metal contacts within .002 inches of each other. The tiniest wires available at the time were three times that size, so they tried a tiny strip of gold foil. It worked. A current came in at one contact point and went out, greatly amplified, from the other contact point.

Bardeen and Brattain's transistor was a success but it was the improvements introduced by William Shockley that made the transistor easy to manufacture. Very quickly, the electronics industry realized that the transistor could not only amplify signals, it could regulate voltage, and switch the course or flow of electricity through a circuit. Since 1947, the transistor has become essential to the circuitry of virtually every electronic device on the market. For their invention of the transistor, Bardeen, Shockley, and Brattain won the 1956 Nobel Prize for Physics.

The Nobel Prize–winning inventors of the transistor:
John Bardeen, William Shockley, and Walter Brattain

FRISBEE®

In the last sixty years, more than 200 million Frisbees have been sold—that's more than the sales of all footballs, baseballs, and basketballs combined. World War II veterans Walter F. Morrison and Warren Franscioni, invented the toy in San Luis Obispo, California. Their inspiration came from kids who tossed metal pie plates to one another, a game the troops had played during the war. But the metal pie plate had its problems: If it was caught improperly, it hurt the catcher's hands, and if the plate struck the ground too hard, it cracked and the game was over. Morrison and Franscioni experimented with different types of plastic that would be durable but wouldn't sting when caught. They came up with the classic design of the rounded edge and marketed their new toy as The Flying Saucer. Neither men knew anything about marketing or distribution, so the toy languished. Eventually they split up (the details of which are a source of acrimonious dispute) and Morrison tried to sell the toy on his own as The Pluto Platter.

Morrison's Pluto Platter was spotted by two toy entrepreneurs, Rich Knerr and A. K. "Spud" Melin, founders of the Wham-O toy company which had introduced the world to the Hula-Hoop. They signed a deal with Morrison, who, within the next few years, received more than $1 million in royalties from sales of the toy. Franscioni, however, received nothing.

To build up demand for the toy, Wham-O's sales representatives visited colleges on the East Coast where they handed out free Pluto Platters to the students. While at Yale, they saw kids tossing metal plates and yelling "Frisbie!" when the plate was in flight. That cry was inspired by the Frisbie Baking Company, whose pies were delivered in metal plates stamped with the Frisbie name. The marketing department at Wham-O liked the name, spelled it Frisbee, and made it the toy's registered trademark.

1948

College students play a game with the precursor to the Frisbee, the pie plate

Zamboni

Throughout the 1920s and 1930s, Frank Zamboni (1901–1988) and his brother Lawrence built large refrigerator units and ran a wholesale block ice business. In 1939, they branched out, opening the Iceland Skating Rink in Paramount, California. A craze for ice skating was sweeping the nation but there were very few ice rinks in southern California. The Zamboni brothers' indoor rink was one of the largest in the United States—20,000 square feet of ice, big enough to accommodate 800 skaters at the same time.

Resurfacing the ice was labor intensive and time consuming. It also cost money—the rink had to be closed to customers while a tractor dragged a scraper over the ice to make the surface even. Three or four workmen followed the scraper, shoveling up the shaved ice, or snow as it was called. Then the whole rink was cleaned and fresh water sprayed over the entire surface, which took about an hour to freeze.

For the next nine years, Frank Zamboni experimented with various machines that would resurface the ice quickly and efficiently. After many false starts, in 1949 he built what he called the "Model A Zamboni Ice Resurfacer." It resembled a Jeep. The Model A had four-wheel drive and four-wheel steering. It was the machine that did everything: It scraped the ice, held the snow in a tank, washed the ice, and applied a fresh sheet of water to freeze. And it required only one man to drive the machine back and forth across the rink. It worked so well that when the famed Olympic skating star, Sonje Henie, saw a Zamboni in action, she ordered two that she could take with on her world ice skating tours. After that, the Ice Capades ordered one, too. Since then, Zambonis have been used to resurface the ice at skating rinks, the Winter Olympics, and countless hockey games. In 2002, Frank Zamboni's invention was named the "Official Ice Resurfacer of the NHL."

The Zamboni is a mainstay of ice rinks everywhere

Artificial Cardiac Pacemaker

In 1949, Dr. John A. Hopps (1919–1998) was studying hypothermia at Canada's National Research Council. As part of a series of experiments on how to restore body temperature, Dr. Hopps tried radio frequency heating. During these experiments, Hopps discovered that if a heart stopped beating due to exposure to very cold temperatures, it could be started again by sending an electric impulse to the heart. The result of Hopps's discovery was the creation of the first artificial cardiac pacemaker. The mechanism placed electrodes on the heart muscles to monitor the normal beating rhythm. If the pacemaker failed to detect a beat at the regular rate, it delivered a quick, low voltage, electric impulse to restore the heart's normal beating pattern.

The early pacemakers were worn outside the body; they were bulky, often uncomfortable for the patient, and had to be replaced frequently. The first completely implanted pacemaker was developed in Sweden by engineer Rune Elmqvist (1906–1996) with surgeon Ake Senning. It was implanted by opening the patient's chest in surgery and attaching the pacemaker to the muscular tissue of the heart, the myocardium.

Today, a typical pacemaker is about 4 centimeters long. The casing is usually constructed of titanium, which in most cases is not rejected by a patient's immune system. The pacemaker itself is hermetically sealed and usually powered by a lithium battery.

New pacemakers continue to be developed. Biventricular pacemakers can monitor both the right and left ventricles of the heart. Heart failure patients often suffer from ventricles that do not beat together; a biventricular pacemaker can synchronize the beating of both ventricles of the heart. There are even pacemakers that can monitor and correct the beating of the ventricles and the contractions of the atria, the blood collection chamber of the heart.

A 1958 pacemaker compared with one from 1978

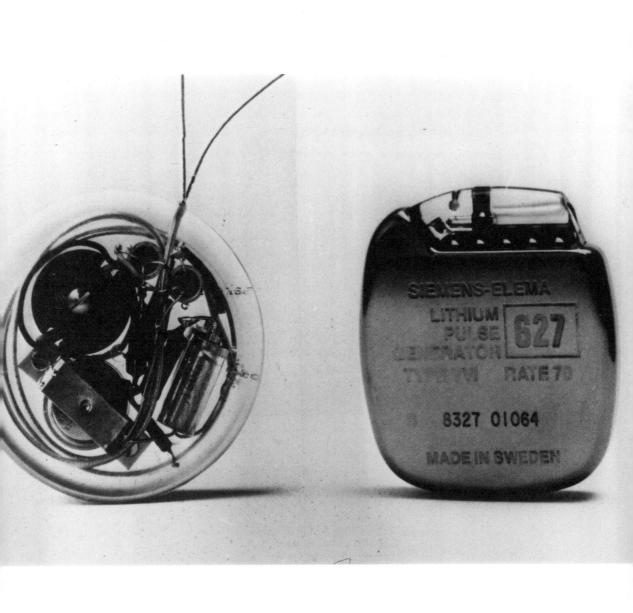

SIEMENS-ELEMA
LITHIUM
PULSE 627
GENERATOR
TYPE VI RATE 70

8327 01064

MADE IN SWEDEN

CREDIT CARD

In the first half of the twentieth century, various stores and even gasoline stations issued credit cards to preferred customers. The merchants' goal was to encourage customer loyalty by granting their customers the convenience of buying now and paying later. It was certainly convenient but also unwieldy. The Marshall Field's credit card was good only at Marshall Field's. The Max's Shell Station credit card was good only at Max's gas station. A shopper who planned to make purchases on credit at many stores was obliged to bring along a thick stack of store-specific credit cards.

In 1949, Frank X. McNamara, head of the Hamilton Credit Corporation, invited two friends to dinner—Alfred Bloomingdale (grandson of the founder of Bloomingdale's department store) and Ralph Sneider, McNamara's attorney. At the end of the evening, when McNamara reached into his pocket to pay, he discovered he had forgotten his wallet. He had to call home and have his wife come to the restaurant with a wad of cash. This embarrassing situation gave McNamara an idea—a credit card that could be used at many locations. He explained his idea to Bloomingdale and Sneider, and together the three men pooled some money and in 1950 started The Diners Club.

The Diners Club was the middleman: It solicited restaurants, hotels, and other businesses to accept its credit card, and it issued a Diners Club card to eligible consumers. Diners Club billed the credit-card holders, then sent payments to the various businesses. Diners Club did not charge its cardholders interest but it made money by collecting a $3 annual membership fee from cardholders, and collecting 7 percent of each transaction made on a Diners card from the businesses. In its first year, Diners Club issued 20,000 credit cards but McNamara figured it was just a fad—in 1952, he sold his share of the company to Bloomingdale and Sneider for $200,000. But the concept of such a credit card caught on. In 1958, Diners Club was joined by American Express and Bank Americard (the forerunner of VISA), and the credit card phenomenon was off and running.

Frank X. McNamara, the inventor of the credit card

DISPOSABLE DIAPERS

Marion O'Brien Donovan (1917–1998) grew up in a family of inventors—her father and her uncle had invented a lathe to grind gears for automobiles. As a little girl, she spent much of her time in her family's machine shop and learned that there were often innovative solutions to everyday problems.

In the 1940s as a wife and mother of small children, she turned her ingenuity to building a better diaper. At the time, diapers were made of cloth and fastened with pins. Every soiled diaper had to be washed, and the soiled sheet the baby had been laying on had to go into the washer, too. Donovan's first invention was a leakproof, waterproof diaper cover. She started by cutting up plastic shower curtains for her prototype but found that nylon worked better and did not give the baby diaper rash. Instead of pins, the diaper cover was fastened with snaps. Donovan's diaper covers debuted at Saks Fifth Avenue in New York in 1949 and were an instant success with moms.

Now that she had waterproofed diapers, Donovan tackled the chore of washing load after load of soiled diapers everyday. In 1950, she developed a disposable paper diaper that drew moisture away from the baby's bottom, once again in the interests of avoiding diaper rash. Incredibly, the idea did not catch on. Ten years would pass before a major U.S. manufacturer, Victor Mills, saw the genius of Donovan's idea and introduced the world to Pampers®.

Once her own children had grown up, Donovan returned to college, earned a degree in architecture from Yale, and designed and built a new house for her family in Greenwich, Connecticut. By the time of her death, she had been awarded a dozen patents.

1950

A woman washes cloth diapers in 1943

Bar Code

ISBN 1-60376-039-3

For many years, two of the biggest headaches suffered by retailers were keeping an accurate record of inventory (all of the items in the store) and preventing shoppers from peeling the price tag off of a low-priced item and sticking it onto a high-priced one. In the late 1940s, the president of a supermarket chain asked a dean at Philadelphia's Drexel Institute of Technology if there was some way to encode product information on each thing in a store, which then could be recorded when the customer purchased the item. The dean replied that it was impossible.

By chance, graduate student Bernard Silver overheard the conversation and repeated it to a friend, Norman Woodland. The two grad students thought it was possible to create such an encoding device. Inspired by Morse code, which uses dots and dashes to transmit information, Silver and Woodland came up with the idea of storing information in parallel lines of varying widths and also in the spaces between these lines.

This bar code would do everything the frustrated president of the supermarket chain hoped for: it could track which items were selling well and which were languishing on the shelves; It could track which items sold better during some seasons and not well during others. Also, since the cost of the item was encoded, there was no price tag for shoppers to fool with.

The tricky part would be inventing an inexpensive, reliable scanner to read the code and record the information. Silver and Woodland built the first scanner themselves; it was not ideal. It was huge, about the size of a desk, and it required a 500-watt bulb to read the bar code. The heat emitted by the bulb was so intense that any paper product, including labels on canned goods, that passed through the scanner would smolder.

Silver died in 1963 before the scanner was perfected. Woodland's solution was a laser scanner that could read the bar code without risk of setting the item on fire. The first laser scanner was installed in a supermarket in Troy, Ohio, and the first product to have its bar code scanned was a package of chewing gum.

An optical scanner reads the barcodes printed on supermarket goods

Fiber Optics

No thicker than a human hair and made of very pure glass, fiber optic strands use light to carry digital information and images over long distances. Hundreds, even thousands, of these strands are bundled together into an optical cable, held together and protected by an outer covering known as the jacket.

The information begins with a transmitter, which produces and sends out the encoded light signals through optic fiber. As pure as the glass of the fiber may be, some impurities do exist and after a distance of half a mile or so, these impurities can break up the light signals; to correct this a laser, known as the optical regenerator, amplifies or strengthens the light signal so it will arrive at its destination clear and intact. The optical receiver decodes the digital information and passes it on to the computer, television, or telephone that is the light signal's ultimate destination.

Fiber optics enjoys a host of advantages over the traditional copper wire used in telecommunications. It is cheaper, lighter, and more flexible than copper wire; it is smaller that copper wire; and it can carry many more signals. Furthermore, fiber optic cable does not degrade as quickly as metal wire; less power is needed to transmit information through it than through copper wire; and since the information is carried via light rather than electricity there is no danger of fire. For these reasons, fiber optics has been adopted by the medical profession for minimally invasive examinations of the inside of a patient body, as well as for what is known as "pinhole surgery."

The man hailed as "the Father of Fiber Optics" is Dr. Narinder Singh Kapany of India, who began his experiments in this new technology while studying optics at the University of London.

Dr. Narinder Singh Kapany, an innovator in the use of fiber optics, demonstrates one of their uses

Calculator

In the early 1970s, electric pocket calculators cost several hundred dollars. By the early 1980s, banks and department stores were giving pocket calculators away as a premium to customers who applied for a credit card.

Although mechanical calculators were developed as early as the seventeenth century in England and France, the first electronic, all-transistor calculator was produced in the United States by IBM in 1954. The machine, the IBM 608, could occupy a small room and cost $80,000. Three years later, the Casio Computer Co. in Japan invented a "compact" calculator about the size of typewriter. In 1964, Sharp developed an "economical" calculator that cost $2,500—but it weighed 55 pounds. The trouble was the state of technology in the 1950s and 1960s: Calculators at that time required several circuit boards and hundreds of transistors to operate.

The invention of microchips in the 1970s transformed the calculator market. Since these new calculators required only a few microchips and a rechargeable battery to operate, they were much more compact than the earlier models. Sharp, Sanyo, and Texas Instruments were just a few of the manufacturers that produced early models of the new handheld calculators. By 1973, algebraic entry calculators were on the market, followed a year later by even more sophisticated calculators that could handle logarithms and perform trigonometry functions.

As calculators became easier and cheaper to produce, many electronics companies got out of the business—with basic four-function calculators selling for just a few dollars, the profit margin had shrunk to next to nothing. Today, the calculators people use most often are found online or in their cell phones.

A mechanical calculator from the 1920's

ROBOT

Robots don't always look like people; mechanical arms on assembly lines are considered robots too

The word *robot* entered the world's vocabulary through a play, *R.U.R. (Rossum's Universal Robots)*, written in 1920 by Karel Capek, a Czech. The story opens in a factory where "artificial people" are manufactured. (Actually, they aren't mechanical robots, they are closer to androids.) The word *robot* comes from *robota*, a Slavic term for serf or slave labor, and it was suggested by Capek's brother Josef.

The notion of a manmade creature—such as the Golem or Frankenstein's monster—is a common motif in both ancient legends and modern science fiction. For the New York World's Fair of 1939, Westinghouse built a robot—7 feet tall, weighing in at 265 pounds—named Elektro. It could walk on command, blow up balloons, move its arms and legs, and "talk" by playing back pre-recorded responses. Aside from a publicity gimmick for the Westinghouse company, Elektro served no practical purpose.

The first useful robot was constructed by George Devol (1921–) of Louisville, Kentucky. Devol called it Unimate and made no attempt to give his robot a humanoid appearance. This was an industrial robot, a machine that could do dangerous tasks more easily and much more safely than human workers. Devol sold Unimate to General Motors who placed the robot in their Trenton, New Jersey, factory where it removed hot metal from a die-casting machine.

The term *robot* is also applied to machines that vacuum a room or mow the lawn without any human pushing them back and forth across the carpet or the yard. The military has remote-operated aircraft that it refers to as robots. Through very powerful long-range remote control systems, these robots can be ordered to scout enemy territory or fire on a designated target.

The type of robot that populates scary science fiction movies was first imagined by Isaac Asimov in his 1942 short story, "Runaround." These robots are programmed to obey three cardinal laws:

1. A robot may not harm a human being or allow a human being to be harmed.
2. A robot must obey a human's orders except when the order conflicts with the First Law.
3. A robot must protect its own existence as long as such protection does not conflict with the First or Second Law.

These robots have not been built . . . yet.

The robot "Elektro," who made his debut at the 1939 World's Fair

SYNTHETIC DIAMOND

In 1893, French chemist Ferdinand Frederick Henri Moissan (1852–1907) claimed that he had created a synthetic diamond by heating charcoal and iron at 7,232 degrees Fahrenheit (4,000 degrees Celsius). For decades afterwards, other scientists tried to replicate Moissan's method but without success. Since he was a distinguished man of science who won a Nobel Prize in 1906, Moissan's colleagues have been at a loss to explain the discrepancy, although it is possible that Moissan's workmen "seeded" the results with fragments of real diamonds so the boss would believe his experiment had been a success.

Whatever may have happened in Moissan's laboratory, the first person to create a synthetic diamond using a process that could be repeated, verified, and in fact was witnessed, was Howard Tracy Hall (1919–). He was working for General Electric at the company's laboratory in Schenectady, New York, when he decided to try to replicate the amount of pressure and heat that carbon experiences before it is transformed into a true diamond. Hall's method required 18 GPa of pressure (which is 180,000 times the pressure of the Earth's atmosphere) and 9,032 degrees Fahrenheit (5,000 degrees Celsius) to dissolve graphite in molten nickel, iron, or cobalt. It created a synthetic diamond 150 micrometers across.

Hall's synthetic diamonds were especially useful as industrial abrasives and General Electric launched GE Superabrasives to market Hall's invention. But in 1970, Hall created a synthetic gem-quality diamond for GE; unlike the tiny industrial diamonds he made in the 1950s, Hall's synthetic gems ranged from 1 to 1.5 carats.

Although Hall's method of creating synthetic diamonds may sound daunting, it is actually cheaper than mining real diamonds, which explains why they have found a wide market in industry and electronics, as well as among consumers who want a large stone for jewelry but cannot afford a true diamond. Karl Marx once wrote, "If we could succeed, at a small expenditure of labor, in converting carbon into diamonds, their value might fall below that of bricks." Synthetic diamonds aren't that cheap but their price is competitive.

A worker manufactures synthetic diamonds

HOVERCRAFT

During World War II, an American, Charles J. Fletcher, invented a vehicle that traveled over water by gliding on air. He called it the Glidemobile. The U.S. Department of War appreciated the possibilities of such a mode of transportation in wartime, and so declared that Fletcher's invention was classified and refused to let him patent it.

Ten years after the war ended, Christopher S. Cockerell (1910–1999), a British engineer, retired from private industry to Norfolk where he and his wife owned and managed a marina. Running the marina must have left Cockerell with some time on his hands because soon he was fooling around with an idea for a machine that would lift ships out of the water and let them skim over the surface on a cushion of air. Freed from having to propel its way through the resistance of the water, such a craft would be able to travel faster than a typical ship—at least that was Cockerell's conjecture.

Experimenting with empty aluminum cans and a vacuum cleaner, Cockerell found that when a smaller can was placed inside a larger one and air was blown out of the small can, it did indeed hover over the surface of the ground and water. He built a prototype in 1955, named it the hovercraft, and received a patent for his invention in 1956. In a demonstration before British military authorities, Cockerell showed how his hovercraft could move effortlessly over water, mud, swamp, and dry land. The British National Research and Development Agency purchased the first hovercraft, and one was built specifically for passengers in 1962.

Contemporary hovercrafts are equipped with a large propeller at the rear or stern that moves it forward. From the bottom of the craft hangs a longish rubber barrier known as a skirt; once the giant fan starts generating air under the craft, the skirt traps it. It is on this air cushion that hovercraft rides to its destination, traveling at a speed of 50 knots (58 miles per hour).

Lord Louis Mountbatten (right) at the inauguration of the hovercraft, with its inventor Christopher Cockerell

Velcro®

Velcro occurs in nature. Well, almost. Some plants have seed sacks (burrs or cockleburs) that attach themselves to the fur of any animal that brushes against them. When its animal carrier dislodges the burrs by scratching or brushing against a harder object such as a tree trunk or a rock, the seed sacks fall to the ground and the plant propagates in a new territory.

In the summer of 1948, a Swiss inventor, George de Mestral (1907–1990), went on a hike with his dog. By the time they returned home, the dog's fur and Mestral's trousers were covered with prickly burrs. In a moment of inspiration, de Mestral pulled off his pants and placed them under his microscope. Through the lens he saw the burr's countless tiny hooks which had caught the threads of the pants fabric. The burrs and the fabric gave de Mestral an idea for a new type of fastener with stiff, tiny hooks on one side and soft loops on the other. When pressed together they would form a strong bond that would not give until the two edges were peeled apart.

With the help of a weaver from a French textile mill, de Mestral developed his new hook-and-loop fastener; he found that nylon when sewn under infrared light formed the tough hooks that would stand up to repeated use. In 1955, de Mestral received a patent for his new fastener, which he called "Velcro," a combination of the French words *velour* (a soft fabric) and *crochet* (hooks).

De Mestral founded Velcro Industries to manufacture and market his new fasteners; within a few years, he was selling more than 60 million yards of Velcro annually.

A seed sack, or "burr," nature's Velcro

Television Remote Control

A reporter once asked Robert Adler (1913–2007) if he felt bad about being single-handedly responsible for giving rise to the fine art of channel surfing and creating millions of couch potatoes. Adler said no. Besides, he added, it is "reasonable and rational to control the TV from where you normally sit and watch television."

Adler's invention, made under the aegis of the Zenith Radio Corporation (which in the 1950s was also manufacturing televisions) was called the Zenith Space Command. Informally among Zenith employees it was known as the "Lazy Bones." Adler's remote control used ultrasound, the high-frequency sound inaudible to most humans (although dogs could hear it), to control the channels and the volume. By pushing one of four buttons on the remote, the user struck one of four lightweight aluminum bars inside the box. Each bar emitted a different frequency that the television recognized. Based upon which frequency was emitted, the TV either changed the channel up or down, or raised or lowered the volume.

Today's remote controls shoot out a beam of light invisible to the human eye. This beam carries the signals of what the operator of the remote wants to do. The TV, CD player, and other electronic appliances detect the light, read the signal, and respond.

Remote controls are not just used for entertainment purposes. For example, robots driven by remote control are used in radioactive or toxic environments where humans cannot or should not go.

It took time for Adler's remote to became a common household device; between its invention in 1956 and the 1980s when remotes became more commonplace, nine million remote controls were sold. Today, however, 100 percent of all televisions, VCRs and DVD players sold in the United States come with a remote control.

A woman demonstrates an early remote control in 1972

SATELLITE

The Russian dog Laika prepares to enter space in *Sputnik II*, the second Russian satellite

The birthday of the Space Age is October 4, 1957, the day the Soviet Union launched the first artificial satellite, *Sputnik I*. Weighing 183 pounds and about the size of basketball, *Sputnik I*'s job was to map the surface of the Earth as it orbited the planet (it took this first satellite about ninety-eight minutes to make a complete circuit around Earth). The United States had also been working on building a satellite, and the fact that the Soviets got into space first caused anxiety if not outright alarm—if the Russians could launch a satellite, did they also have the technology to launch intercontinental nuclear missiles? Before the Americans could catch their breath, the Soviets launched another satellite on November 3—*Sputnik II*. This one carried a dog named Laika (which the American press nicknamed "Muttnik").

During the Cold War, satellites seemed ideal for pinpointing an enemy's secret military installations and missile silos. Satellites are still used for this purpose but by and large their most common uses are infinitely more benign. Satellites handle telephone calls, transmit television stations via DIRECTV and the DISH Network, track weather systems, and give drivers directions via a Global Positioning System. Satellites also process distress calls from downed aircraft, damaged ships, or even individuals who are lost or injured in remote locations. Satellites do this by receiving and transmitting radio signals.

Throughout their history, satellites have been custom-designed for a specific purpose; the GPS satellites are the exception—at least twenty identical such satellites are orbiting Earth at this moment. In addition to satellites that beam back to Earth all types of useful information, there are other satellites, known as "space junk," still circling the planet. These satellites have either expired or are obsolete. The *Sputniks* are not floating out there amid the space junk—they fell back toward the planet and were burned up as they re-entered the atmosphere.

A satellite orbiting Earth

LASER

The term *laser* is so familiar to us that we've forgotten that it is an invented word—an acronym for Light Amplification by the Stimulated Emission of Radiation. And before there was a laser there was a maser. Arthur L. Schawlow, a Bell Labs researcher, and Charles H. Townes, a consultant to Bell Labs, invented a Microwave Amplification by Stimulated Emission of Radiation (maser) in 1954, which was similar to a laser but did not use a beam of light that could be seen by the naked eye. In 1958, however, Schawlow and Townes speculated that it was possible to create a laser. And they were correct—in fact, two other research scientists were already developing a laser. Gordon Gould, a student of Charles Townes at Columbia University, generated the first optical laser in 1958 but did not file for a patent until 1959. By that time, Theodore Maiman had already patented his laser. Both Gould and Maiman had discovered how to control and concentrate the power of light.

A laser controls the way energized (also known as "excited") atoms release light energy or photons. For a laser to work it must have a large number of excited atoms; atoms become excited by repeated bursts of intense light or repeated bursts of electrical discharges.

A laser differs from normal light. For example, unlike a flashlight or lamp that sends out light in every direction and therefore is rather weak, a laser sends out a single, strongly concentrated beam. In normal light, you'll find the entire spectrum of colors; a laser, on the other hand, is only one color because it sends out one specific wavelength of light.

Because lasers are so sharply focused, they can be used for a wide variety of precision work, from surgery and playing a CD to reading barcodes at the checkout counter and removing tattoos.

The use of lasers to perform precise corrective eye surgery has become the norm

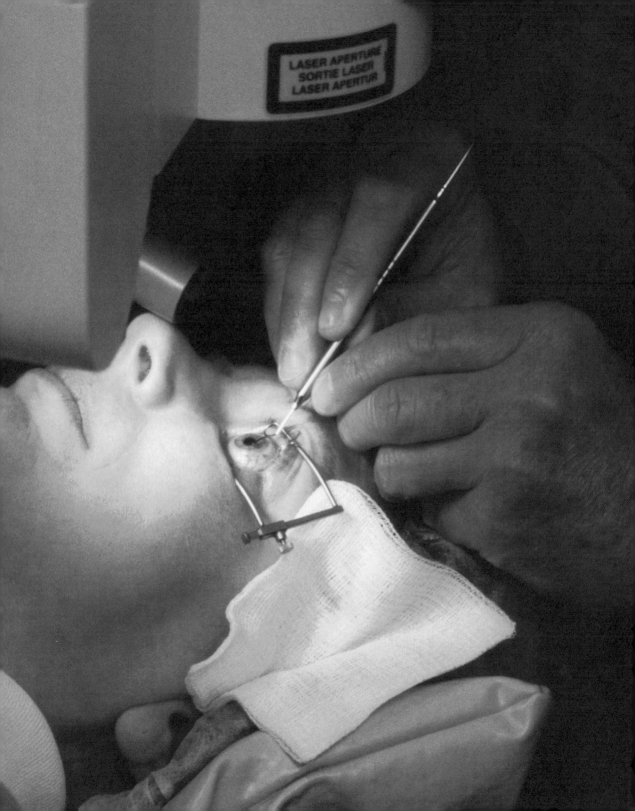

THREE-POINT SEAT BELT

One of the least popular options on automobiles during the 1950s was the seat belt that buckled across the driver and passengers' laps. The two most common complaints were the belt was confining—and, it wrinkled the driver's and passengers' clothes. As a security device, the lap belts were inadequate: They did hold people in place but the force of impact in a car crash could still propel the upper body forward against the steering wheel or dashboard, resulting in severe injuries.

In Sweden, inventor Nils Bohlin was designing a new type of seat belt that would be a better safety measure than the lap belt. He took his ideas to Gunnar Engellau, president of AB Volvo; as it happened, Engellau had lost a member of his family in a car crash, so he was receptive to Bohlin's ideas for a better safety belt. Engellau gave Bohlin a research and development job at Volvo so he could continue his work.

Drawing upon the type of safety belts he had designed for aircraft pilots, Bohlin designed a three-point seat belt that ran from one shoulder, across the chest, and across the hips. Unlike the old lap belt, Bohlin's new belt would prevent the upper body from pitching forward in an accident. In 1958, Bohlin patented his three-point seat belt; in 1959, Volvo installed it in all of its cars; and in 1963, Volvo brought the three-point seat belt to America. Today, Bohlin's seat belt is standard equipment in virtually every car and truck.

Nonetheless, it took time for the use of seat belts to gain acceptance among American drivers. According to the National Highway Traffic Safety Administration (NHTSA), as late as 1984, only 14 percent of drivers and passengers in the United States wore their seat belts; by 2004, 80 percent of drivers and passengers were wearing seat belts. One reason for the increase in usage is that virtually every state requires seat belts to be worn but many drivers and passengers have also become convinced of the value of wearing seat belts. NHTSA statistics show that wearing a three-point seat belt reduces fatalities in car accidents by 45 percent and reduces fatalities by 60 percent in light truck accidents.

The three-point safety belt has proven to be much safer than an across the waist belt

Ultrasound Imaging

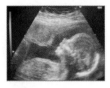

The term *ultrasound* has only one meaning for expectant parents—a blurry but marvelous photograph of their developing baby in the womb. Yet very few happy families realize that the miracle of ultrasound emerged out of the wartime technology of radar and sonar.

The man primarily responsible for the use of ultrasound as a medical diagnostic tool is Dr. Ian Donald (1910–1987), a Scottish physician. He served in the Royal Air Force during World War II, where he learned how radar was used to detect incoming enemy planes and missiles, and sonar was used to detect enemy submarines. After the war, Dr. Donald was appointed to the Regius Chair of Midwifery at the University of Glasgow, where he pursued his interest in using sonar to locate fibroid tumors and ovarian cysts in women patients. He called his new method ultrasound. It was essentially a scanning technique that sent out inaudible sound waves through the patient's abdomen; when they struck a solid mass such as a tumor or cyst, the sound waves bounced back and an image of the mass was transmitted to a computer screen. Dr. Donald explained his method in a groundbreaking article published in Great Britain's most prestigious medical journal, *The Lancet*, in 1958.

From scanning an abdomen for tumors and cysts, it was only one step to scanning an abdomen for a baby. The first ultrasound image of an infant in utero appeared on Dr. Donald's computer screen in 1959. In the years that followed, he and his colleagues used ultrasound to chart fetal development and found that they could identify a host of complications that could affect the health of the baby and or the mother.

In common parlance, *ultrasound* continues to mean "pictures of the baby" but it has many more medical uses. It can monitor blood flow through the heart, kidneys, and major arteries; spot kidney stones; detect an early stage of prostate cancer; and identify any abnormalities inside the heart. It is one of the many valuable diagnostic tools medicine has today.

A pregnant woman receives an sonogram

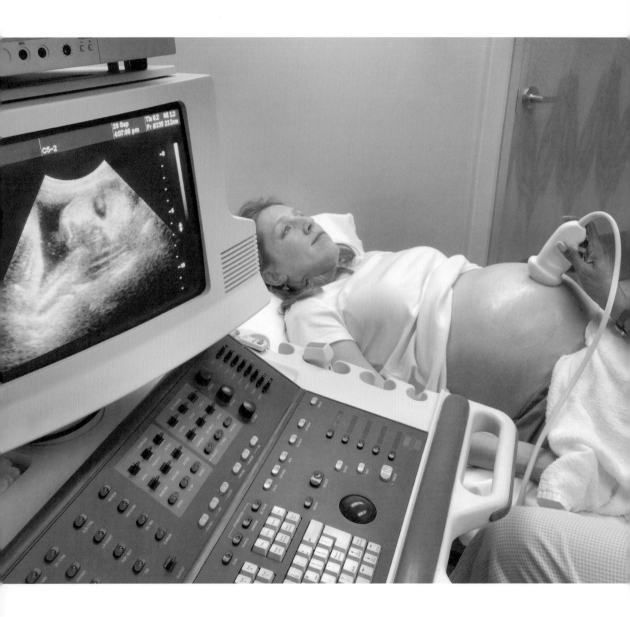

In Vitro Fertilization

The first successful experiment of in vitro fertilization did not produce a test tube baby but a litter of test tube rabbits. The man responsible was Dr. Min Chueh Chang (1908–1991), a Chinese-born reproductive biologist who had immigrated into the United States. Dr. Gregory G. Pincus (1903–1967) had claimed that in 1935 he had achieved the first successful in vitro fertilization that resulted in the birth of a live rabbit, but no one had been able to replicate his experiment—a situation which raised doubts in the scientific community. For his experiment, Dr. Chang fertilized the eggs of a black rabbit with the sperm of a black rabbit and then implanted the fertilized eggs into a white rabbit. The white rabbit delivered a litter of tiny black bunnies. Dr. Chang and his colleagues went on to replicate this experiment with other small mammals such as hamsters and mice.

In vitro is a Latin phrase that means "in glass," and refers to such glass equipment as Petri dishes and test tubes that scientists used in laboratories in the 1950s. Since eggs and sperm were brought together in these glass vessels, the method became known as in vitro fertilization, or IVF. The purpose of IVF was to assist couples who had fertility problems and could not conceive a child in the usual way. Nonetheless, nineteen years would pass between Dr. Chang's experiment with black rabbits and the birth of the first "test tube baby."

The men responsible were Patrick C. Steptoe (1913–1988), a British obstetrician and gynecologist, and Robert G. Edwards (1925–), a British physiologist. They had been consulted by a British couple, Leslie and John Brown, who had been told by other doctors that they could not have children because of a blockage in Leslie Brown's fallopian tubes. In their laboratory, Steptoe and Edwards fertilized one of Leslie Brown's eggs with her husband's sperm, then implanted it in Leslie's uterus. On July 25, 1978, Leslie Brown gave birth to a healthy 5-pound, 12-ounce baby girl; her parents named her Louise. Later the Browns' returned for another round of IVF and had another daughter, Natalie.

The first "test tube" baby with in vitro pioneers Dr. Robert Edwards and Dr. Patrick Steptoe

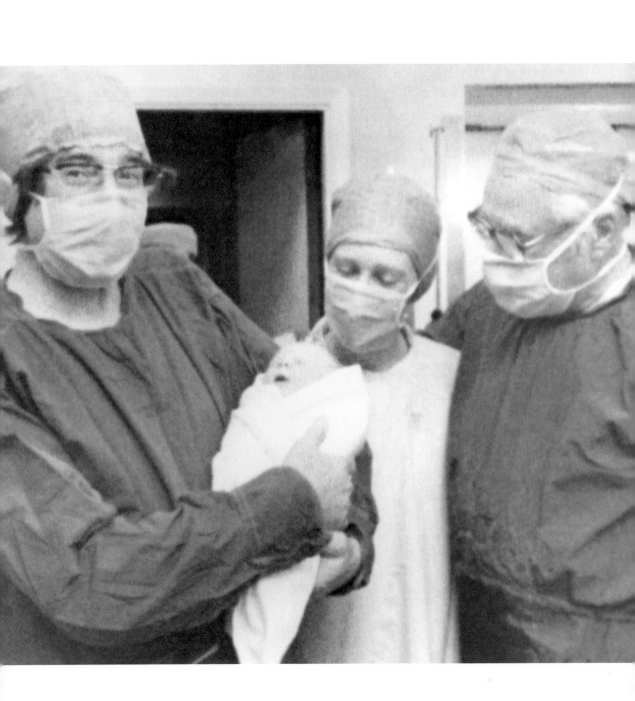

ARTIFICIAL TURF

The opening of the Houston Astrodome in 1965 was a public relations masterpiece. President Lyndon B. Johnson and astronaut Gus Grissom led the list of distinguished guests; attractive young women in gold lamé miniskirts and blue go-go boots escorted fans to their seats, while on the field the grounds crew—renamed the Earthmen—tidied up the field in orange space suits. Best of all was the pleasure of watching a ball game in a climate controlled indoor stadium. Houston's hot, humid weather and jumbo-sized mosquitoes had made going to a ballgame an ordeal. It was not unusual for dozens of fans to be treated for heat exhaustion, and the concession stands did a booming business selling insect repellant.

Great as it was, one thing had not been factored into the design of the Astrodome—the glare of the sunlight reflecting off the dome's glass panels. Players lost sight of fly balls in the glare. To solve the glare problem in the new stadium, Judge Roy Hofheinz, who owned the Houston Astros, had a coat of gray paint applied to the dome's panels: it eliminated the glare but killed the grass. The sleek, futuristic Astrodome had become a dust bowl.

Fortunately, a year earlier, the chemical company Monsanto had delivered a new artificial fabric that looked like a lovingly tended lawn: the name of the new fabric was ChemGrass. It was green, just like real grass; it had individual strands just like real grass; and balls bounced on it, just like on real grass. Hofheinz placed an order with Monsanto to cover the barren dusty field with ChemGrass. On March 19, 1966, at an exhibition game, the public saw their first ball field covered in artificial turf—AstroTurf, the sports writers called it. It looked gorgeous and it did not need sunlight, water, weed-killer, or fertilizer.

Today corporate parks, private homes, city parks, and even cemeteries have installed artificial turf.

The Houston Astrodome, the first athletic field to use astroturf

ATHLETIC SHOE

Athletes at the ancient Olympic Games in Greece competed barefooted (and naked). It was not until 1917, that a shoe was marketed directly to athletes. The shoe's brand name was Keds® and it had a rubber sole and a canvas body. The company's sales force promised amateur and professional athletes that this shoe would let them "run their fastest and jump their highest." In fact, these first sneakers were heavy and cumbersome.

In 1958, Phil Knight, a member of the University of Oregon track team, complained to his coach, Bill Bowerman, about his running shoes. That complaint evolved into an athletic shoe empire. By 1964, Knight and Bowerman had designed and manufactured a new lightweight, high-tech, but inexpensive running shoe that they sold personally at track meets around the country. In 1968, they formed a corporation, NIKE, Inc., named for the Greek goddess of victory.

The two men were a great team: Bowerman designed the shoes and Knight marketed them. It was Bowerman who came up with the idea that a waffle sole would give athletes better traction (he fashioned the first waffle sole by placing a piece of rubber in the wafflemaker in his kitchen). Knight's marketing savvy led him to announce in 1972 that "four of the top seven finishers" in the Olympic Trials for the marathon event had won wearing NIKE (he omitted to mention that the runners who took the top three spots had worn Adidas®).

The boom in the fitness industry that began in the 1970s made NIKE a household word, and the company's market share was enhanced by it's catchy "Just do it" advertising campaign, as well as endorsements from superstar athletes such as Michael Jordan. NIKE's distinctive "swoosh" graphic was designed in 1971 by Carolyn Davidson for a fee of $35.

Today, there are many brands of athletic shoes to choose from. In the United States, the athletic shoe industry generates more $13 billion and sells more than 350 million pairs of shoes each year.

Phil Knight, co-inventor of the NIKE athletic shoe

Kevlar®

In 1946, twenty-three-year-old Stephanie Kwolek (1923–) graduated with a degree in chemistry from Margaret Morrison Carnegie College (the women's division of Carnegie Mellon University). She wanted to go on to medical school but she didn't have the money, so she took a job in the synthetic fibers division of Du Pont. The company was the leader in creating synthetic fibers, having already brought nylon, Dacron polyester, and Lycra to the market. Kwolek worked with aromatic polyamides, known as aramids for short. Aramids are characterized by their tight, rodlike molecules—in stark contrast to the flexible chains of molecules that Du Pont had been able to make into nylon.

Kwolek experimented with various solvents to find one that would blend with the aramid molecules to form a polymer. When she finally got a polymer it was cloudy and watery, not clear and thick like the polymers that became Du Pont's most successful synthetics. The next step would be running her polymer through the "spinner" to see if it could be spun into synthetic thread. The chemist who ran the machine looked at Kwolek's goop and refused to test it but Kwolek persisted, and eventually persuaded him to give her unpromising polymer a trial run. To their mutual surprise, Kwolek's polymer was easy to spin into thread and the thread it produced was light but extremely strong.

As she studied her new fiber, she learned that it was five times stronger than steel of the same weight. Kwolek's discovery went on the market in 1971 under the name of Kevlar®. Today, Kevlar is used in everything from bridge cables to fishing line, but it is most famously used as lightweight body armor.

Body armor made of Kevlar is a necessity for police officers and soldiers

Automated Teller Machine

People of a certain age will recall when all bank transactions had to be done by hand, with paper, and in person. This often required waiting on long lines outside a bank teller's window. It was time-consuming, inefficient, and often frustrating—for both the tellers and the customers. In the 1930s, an Armenian immigrant to the United States, Luther George Simjian (1905–1997), invented a cash-dispensing machine, which City Bank of New York installed in its lobby. It flopped. Customers didn't trust the newfangled gadget to give them the correct amount of cash, and after six months City Bank's executives cancelled the experiment.

The idea got a second life in England in the 1960s when a Scotsman, John Shepherd-Barron (1925–), developed the next incarnation of cash-dispensing machines. In an effort to limit the possibility of fraudulent withdrawals, Shepherd-Barron developed paper vouchers encoded with the customer's PIN (personal identification number). The customer would slip the voucher into the machine and then enter the PIN on the machine's keypad. For security purposes, Shepherd-Barron had wanted PINs to be six-digits but his wife complained that it was too hard to remember such a string of numbers, so Shepherd-Barron reduced the PIN to four digits. Shepherd-Barron's ATM was installed in the Enfield Town branch of Barclays Bank in North London. To promote the new service, the bankers arranged for the popular British comic actor, Reg Varney (1916–), to be the machine's first customer.

Initially, the only task the machines performed was to dispense cash. For all other banking transactions, customers still had to go the teller. Such cash-only ATM machines are still found today in such places as shopping malls, airport terminals, and convenience stores. ATMs located on a bank's premises offer many more options, from accepting deposits and payments to transferring funds between accounts.

The first ATMs could only dispense cash to customers who already had an account with the ATM's bank. The 1980s saw the growth of interbank networks, which permitted customers to withdraw cash from any bank's ATM—for a fee.

A woman withdraws cash from the world's first ATM

Air Bags

In the 1950s, two inventors interested in automobile safety, Walter Linderer of Germany and John Hedrik of the United States, developed a safety cushion that inflated with compressed air. The system had two disadvantages: It was not automated—the driver had to trigger it—and the compressed air did not inflate the cushion fast enough to provide good protection for the driver.

In 1968, Allen K. Breed, an entrepreneur and inventor, invented an air bag that deployed and inflated instantaneously with nitrogen gas upon violent impact. In the early 1970s, American automakers began offering air bags as a option. Some models of cars offered driver and passenger air bags, while other models included side air bags, too. There were problems with the early air bags since they tended to hit small adults and children in the face rather than in the chest, causing injuries and even fatalities. Even today, the deployment of air bags injures some passengers and is considered dangerous for small children, so some cars with air bags allow the system to be disengaged.

Advances in air bag design now can take into account the physical size of the driver and the passenger, the position of their seat, if they are wearing a seat belt, and can even gauge the severity of the crash.

1968

A 1970 crash test using air bags

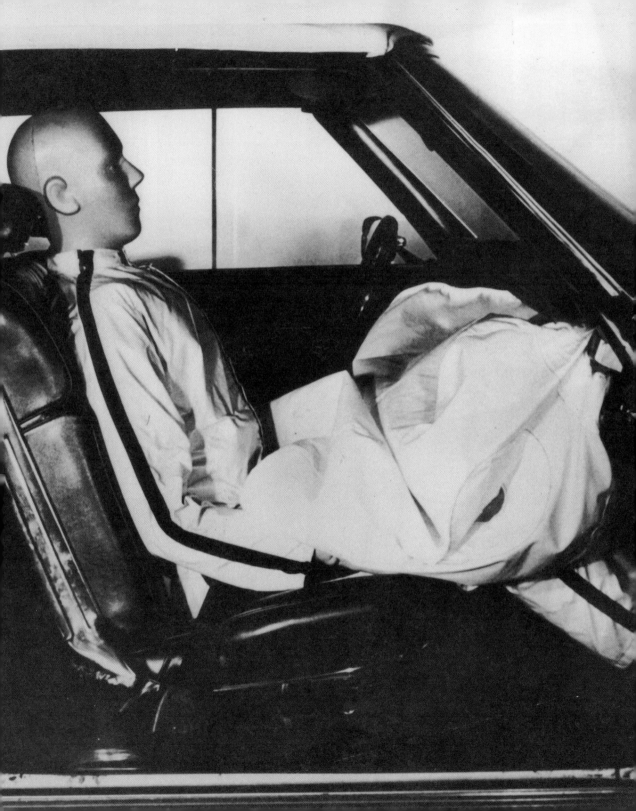

COMPUTER MOUSE

During World War II, Douglas Engelbart (1925–) enlisted in the U.S. Navy and was shipped off to the Philippines to serve as an electronic/radar technician. After the war, he studied for a Ph.D. in electrical engineering at the University of California, Berkeley, then took a job at the Stanford Research Institute. His first two years at SRI were incredibly productive—he was awarded a dozen patents in the field of miniaturization, digital devices, and magnetic computer components.

In 1968, Engelbart signed up as an exhibitor at the Fall Joint Computer Conference in San Francisco where he showcased two astonishing devices. The first was what he called NLS (for No Line System), which used a computer mainframe for video teleconferencing. The second was a gadget for navigating across the screen of the computer monitor. Engelbart called it an X-Y Position Indicator for a Display System. Today, we call it a computer mouse. Engelbart's first mouse rolled about on small wheels; today the mouse rolls over the desk surface with a small tracking ball or relays its position via laser.

The mouse Engelbart introduced was bulky—about the size of an adult's hand. He imagined that computer users would move the mouse around with one hand, while typing on the computer keyboard with the other. He received a patent for his device in 1970 but it remained a high-tech curiosity until 1984 when Apple Macintosh began marketing computers with a mouse. The Mac's mouse was smaller, easier to maneuver but still boxy. Today, computer mice come in a variety of styles and designs—some of the most attractive are designed specifically for gaming, with additional buttons and directional aiming capabilities.

The mouse operates by sensing motion and clicks, converting them into signals the computer can read, and then transmitting that information to the computer, which responds almost instantaneously.

Douglas Engelbart with the first computer mouse

SMOKE DETECTOR

For about $10, you could save your life and your property. That's what a smoke detector costs at any hardware store. Rarely in human history has such a small financial investment produced such great returns—it is estimated that thousands of lives and billions of dollars in property are saved every year thanks to smoke detectors.

There are two types of smoke detectors. The inside of a photoelectric detector looks like a T. Beams of light shoot across the crossbar; the detector is located at the bottom of the vertical bar. When smoke enters the crossbar chamber, the solid particles in the smoke deflect the beams of light downward into the vertical bar. The beams strike the detector and the alarm goes off.

The second and more common type of smoke detector is the ionization model. Inside the ionization chamber of the smoke detector are two plates and very tiny bit—about 1/5,000th of a gram—of a radioactive element known as americium-241. Americium "ionizes" oxygen and nitrogen atoms in the air inside the smoke detector (*ionize* means to knock an electron off an atom). The knocked off, or free, electron has a negative charge. The atom missing one electron has a positive charge. The negative electron is attracted to the positive voltage plate; the positive atom is attracted to the negative voltage plate. The smoke detector senses the small electric current this attraction creates. When smoke enters the ionization chamber, it interrupts the electrical current; the smoke detector registers the interruption and sets off the alarm.

Compact Disk

At age six, James T. Russell (1931–) built a model battleship that he operated by remote control. He personalized the model by adding a compartment to hold his lunch.

In 1965, Russell joined Batelle Memorial Institute in Richland, Washington, as a senior scientist. Russell's passion project at the time was finding a better way to record and play music. Like all music lovers of the time, he was continually frustrated by vinyl LPs that could become nicked, scratched, or warped. He was also not wild about the sound quality of the typical LP. Encouraged by senior management at Batelle, Russell spent the next five years developing a better device to record and replay sounds.

Since the stylus and the turntable were often the problem, he imagined a mechanism in which none of the parts touched—and the only way that would work, Russell, realized, was if he used light. On a photosensitive disk, he implanted microscopic bits or bumps, each one a micron in diameter. A laser read the patterns of the bumps, a computer converted the data into an electronic signal, which in turn was converted into an audio or visual transmission.

Russell found that by making the binary code—the pattern of bumps on the CD—so compact, he could load volumes of information on a single disc.

Finding investors who could see the possibilities of this breakthrough technology was tough, so throughout the 1970s Russell improved and refined his invention. By 1985, he had been awarded twenty-six patents related to the technology of compact discs.

Today, compact discs carry music, movies, and whole libraries of information. Russell's invention has transformed how the world stores and retrieves information. He has been quoted as saying, "I've got hundreds of ideas stacked up—many of them worth more than the compact disc. But I haven't been able to work on them."

A demonstration by Philips in 1981 of the compact disk

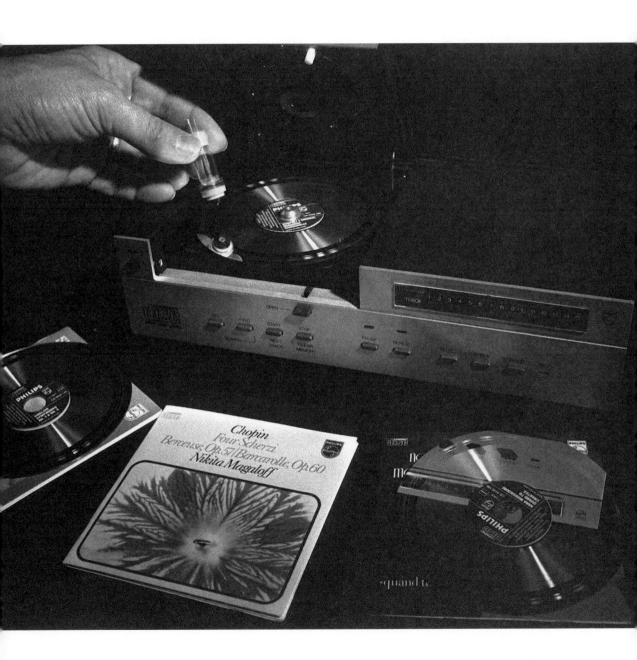

EMAIL

Thomas Edison worked his way through hundreds of possible filaments before he found one that would work in his electric light bulb but email was invented in less than a day. Raymond Tomlinson (1941–), who worked on ARPANET, the precursor of the Internet, says developing a working email system was simple. "It probably took four, five, six hours to do," he said in a 2002 interview. In 1971, a person could leave a message or memo to himself on his own computer but Tomlinson wanted to be able to send messages to other computers. "It would be like the telephone," he said, "but they wouldn't have to be there to answer the phone." In other words, Tomlinson's system could store messages until the computer user was available to read them.

By the way, it was Tomlinson who started using the @ symbol to separate a user's name and the name of his or her computer network.

Tomlinson's colleagues saw the possibilities of the new system immediately but no one called it "email." They called it either a message or just plain mail. But while the new method for sending messages won nearly universal acceptance among computer professionals, the industry was so small and there were so few computers that Tomlinson's invention did not venture outside the world of high technology research and development. Only when personal computers for home and business gained momentum did the idea of email become more applicable.

Nowadays, there are still a few bugs to work out. First is the privacy issue, since email is rarely encrypted to protect the confidentiality of the sender or the receiver. Next is spam, the unwanted, unsolicited email messages that clog up an email inbox (although email software is getting progressively better at screening out spam). And there is always the issue of email etiquette, whether it's writers who send out rude or hostile messages or employees who copy their emails to everyone in the corporation.

Nonetheless, email has changed business and personal communication in ways Tomlinson probably never imagined. When he had the idea back in 1971, he said he thought it would be "a neat thing to try out."

The reliance upon email has led to new businesses, such as the "cybercafe" pictured

SPACE SHUTTLE

At the beginning of the U.S. space program, Maxime A. Faget (1921–2004), an immigrant from British Honduras, was hired by NASA in 1958 to design the *Mercury* capsule. From 1958, until his retirement from NASA in 1981, Faget contributed to the design of every manned spacecraft. Among his achievements is the fact that in 1972, he patented his design for a reusable aircraft for space travel, better known as the space shuttle.

The space shuttle is the world's first reusable spacecraft. With a crew of between five and seven astronauts, it is responsible for crew rotation at the International Space Station, it picks up obsolete or defective satellites, and it can launch new satellites into Earth's orbit. In fact, a space shuttle can carry up to 50,000 pounds of equipment (also known as the payload).

The space shuttle has three main components: the winged aircraft that looks like an enormous fighter jet is known as the orbiter; the crew and the payload are located in the orbiter. The orbiter is mounted on a huge rust-colored external tank that provides the liquid hydrogen fuel at liftoff to the shuttle's three main engines. This part of the shuttle is not reusable, so at the time the shuttle re-enters Earth's atmosphere the tank is jettisoned; this is precisely timed so the tank will break up harmlessly somewhere over the Pacific or Indian Ocean. On either side of the external tank are two slender white rockets, the solid rocket boosters; they provide 83 percent of the thrust at liftoff. These rockets also bear the weight of the external tank and the orbiter. The rockets are jettisoned at liftoff but they can be recovered and used again. With its rockets and external tank gone, the orbiter becomes a glider when it returns to Earth, making what is known as a "dead stick landing" or unpowered landing.

Since 1977, six space shuttles have been built, two of which ended in tragedy: the *Challenger* in 1986 and the *Columbia* in 2003—in both cases no crew members survived. NASA expects to retire the space shuttle in 2010, and introduce a new space vehicle, the *Orion*, to carry crewmembers back to the Moon.

The maiden launch of the space shuttle *Columbia*

EARLY PREGNANCY TEST

It's technical name is qualitative urine human chorionic gonadotropin (HCG) analysis—which explains why it is commonly known by the cozier and decidedly friendlier name, the home or early pregnancy test.

The U.S. Food and Drug Administration (FDA) approved early pregnancy test kits for consumer use in 1976, and the first nonprescription e.p.t. kit went on the market the following year, manufactured by Warner-Chilcott. Competing products—such as ACU, Answer, and Predictor—appeared on drugstore shelves soon thereafter. They all cost about $10 and took about two hours to register a result. They were also a touch unreliable—some women who received a negative reading from their early pregnancy test kit discovered later that they were in fact pregnant. Even today, when the accuracy level of the home test kits has improved, manufacturers still recommend that women who have tested negative should test themselves again in another week or so.

A home pregnancy test is a strip of paper that has been treated with a chemical to detect in urine the presence of HCG, the hormone produced by the placenta soon after an egg has been fertilized. To use a typical home kit, a woman places a test strip in a urine sample or under her urine stream. If a thin blue line or a plus sign appears on the paper strip, the woman is pregnant. There are also home test kits that provide a digital readout.

In some cases, a woman's HCG level may be low, or perhaps at the time she took the test the embryo had not been implanted yet. That is why if the test registers negative but the woman has reason to believe that she is pregnant it is recommended to retake the test five to ten days later. And of course for a truly accurate reading, it's best to schedule a visit with the doctor.

Before home pregnancy tests, women received the news from their doctors

Personal Computer

The story you've heard is true. In the mid-1970s, Steve Wozniak and Steve Jobs, working in Jobs's parents garage in what is now Silicon Valley, California, built the first personal computer—the Apple II. The computer of the two Steves offered consumers expanded memory, data storage, modestly priced programs, a keyboard, and color graphics—all for $1290.

The success of the Apple II all but compelled IBM to come out with their own model—the IBM PC. It had a 16-bit microprocessor and a standardized operating system. Apple employees responded in 1984 by introducing models that were user-friendly: Their innovation was presenting the user with an array of little pictures or icons that represented the computer's various functions. All an Apple user had to do was click on the icon and the computer responded; IBM PC users had to type in commands.

Since then, the evolution of personal computers has moved at breakneck speed. By the late 1990s, computer users were surfing the Web, sending and receiving email, running complex software programs, and playing highly stylized computer games. The introduction of laptops—small, portable computers that users could carry and use anywhere—opened up a whole new market.

1977

Apple founder Steve Jobs pictured in 1984 with the Apple Macintosh, one of the first personal computers

Post-it® Notes

In the early 1970s, Arthur Fry (1931–) of Minnesota worked as a new product development researcher at 3M. He also sang in his church choir. While he loved the music and the camaraderie of the being a choir member, finding his place in the hymnal was a source of constant frustration for Fry. Bookmarks fell out, paper clips and other such office supplies damaged the hymnal's pages. He needed something that would stick to the hymnal pages without spoiling them.

At 3M, Fry had attended a seminar presented by his colleague Spencer Silver (1941–). Silver had developed an adhesive that when applied to a surface was strong enough to hold a piece of paper in place. Silver imagined this adhesive could be sprayed on bulletin boards. The adhesive was strong enough to hold the notices, but once the notice was out of date it could be peeled off the board easily and a new notice posted in its place—and there would be no need to spray another application of the adhesive.

Fry asked Silver for a sample of the adhesive. He coated one side of small strips of paper. When he stuck these strips to another piece of paper, the adhesive held; and when Fry peeled off the strip of paper it did not harm to the page nor did it leave behind any residue. Fry used his sticky slips of paper to mark the pages in his hymnal, and passed them around the office where his co-workers recognized the benefit of this innovation immediately.

3M brought out the new product in 1980, after trademarking the name Post-it® Notes and the product's original canary yellow color (although the notes now come in a rainbow of colors).

Fry and Silver went on to win 3M's highest awards for research and development, as well as a host of international awards for innovation and engineering.

While Post-it notes are useful at the office, they can also be used decoratively

Global Positioning System

Global Positioning System (GPS), the foolproof method for finding your way through an unfamiliar town, has it origins in the development of radar during World War II. The Allies used radar to detect incoming missiles and enemy aircraft; in fact, radar is credited with enabling the Allies to destroy perhaps as much as 95 percent of all V-I rockets the Nazis fired at London. One of the men working on the radar program during the war was Ivan Getting (1912–).

By the 1960s, Getting had started his own company, The Aerospace Corporation, which designed new antiaircraft technology for the U.S. Department of Defense (DOD). It was Getting who suggested using satellites in space to track the movement, and even navigate, vehicles or other objects moving on Earth. In other words, a GPS could identify the path of incoming missiles or artillery shells; guide the flight path of U.S. missiles; lead troops through unknown territory, even in the dark of night; and assist in rescue missions by pinpointing the location of downed pilots.

In 1972, the DOD hired Bradford Parkinson (1935–) to take Getting's ideas for a global positioning system and make them a reality. By 1978, Parkinson had virtually perfected the first GPS program—it was accurate to within 10 feet. (Today it is accurate to within 3 feet.) For his achievement the DOD presented Parkinson with its Superior Performance Award.

It became obvious very quickly that the GPS also had civilian uses—navigation, of course, but also to locate stranded motorists and backcountry hikers.

By the year 2000, GPS systems had become so affordable that they appeared regularly in cars, boats, and even in handheld models. And Ivan Getting was not forgotten—in 2003, in recognition of developing new technology that improves day-to-day life, Getting and Parkinson were awarded the Charles Stark Draper Prize of $500,000.

A soldier uses a handheld GPS device

The Internet

Throughout the course of history, humans have set up networks of roads, water, and sewer systems and telegraph, telephone, and electrical power lines, but they seem modest compared to the vast system of computer networks known as the Internet. It is an enormous global system of thousands of individual computers and computer networks, all linked to each other and able to speak to each to convey information almost instantaneously (assuming the user has finally retired his or her dial-up modem).

The idea for the Internet goes back at least to the 1960s, and in 1973 the design for such a network was complete—although it took another decade before it was ready for use. The Internet links many smaller networks of computers. In each such network, there is a primary computer known as the gateway. All gateways are linked by telephone lines, fiber optic cables, or radio signals. And an infinite number of new gateways can be added. There is no central computer system, no "home office of the Internet," just this endless series of linked computer systems. When you perform a search on your home or office computer, the signal leaves your gateway and runs through all the gateways until it finds what you are looking for and comes back with the link, or a list of possible links. Today, all of this can happen in the blink of an eye.

Among the heroes of this story are Timothy Berners-Lee (1955–), who invented the World Wide Web as a kind of high-tech library where researchers could share information. Building upon the work of Vannevar Bush, Ted Nelson, and Douglas Englebart, he developed the Hypertext Transfer Protocol (HTTP), the language that all the computer networks of the Internet could understand. Berners-Lee also invented the URL, or Uniform Resource Locator, which gave each document on the Internet its own address so it could be found more easily.

Marc Andreessen developed the first truly effective browser, Netscape Navigator. It was also made Andreessen one of the first Internet moguls: The initial public offering of Netscape stock in 1995 made him a multimillionaire overnight.

Timothy Berners-Lee, the foremost Internet pioneer

DNA Fingerprinting

As a boy in England, Alec Jeffreys (1950–) amazed his friends and impressed his teachers with his ability to perform fairly complicated experiments with a common, storebought chemistry set. Those early days of fun with test tubes evolved into a doctorate in biochemistry from Oxford University and a postdoctoral fellowship to study genetics at the University of Amsterdam. The shift to genetics was key to the future of Jeffreys's career. Once he completed his work in The Netherlands, he returned to England as a lecturer and researcher at the University of Leicester. There he studied DNA variation and gene families with an eye toward tracking hereditary diseases.

Jeffreys and his fellow researchers concentrated on studying the strands of DNA known as mini-satellites where genetic variations were easiest to spot. Part of the project was coming up with a process that would enable geneticists to isolate these mini-satellites. Jeffreys had taken X-ray photos of DNA strands from various individuals and on September 10, 1984, was studying the X-ray film when he noticed that the DNA mini-satellites were all different, suggesting that each person's DNA is utterly unique—just like his or her fingerprints. Since DNA can be collected from hair, skin, blood, and other bodily fluids, the legal community in particular was instantly fascinated with Jeffreys' discovery.

The first legal case in which DNA evidence was introduced in court involved a Ghanaian boy who claimed that his parents were British citizens. Immigration officials did not believe the boy and denied him entry to the United Kingdom. The boy's lawyers had a laboratory analyze a DNA sample taken from their client. The results from the lab test proved conclusively that the boy was truly the son of the Ghanaian couple with British citizenship.

In recognition of his extraordinary discovery, Jeffreys has been showered with honors, including election as a Fellow of the Royal Society in 1986. The following year, the Royal Society presented him with their highest honor for advancing the study of chemistry, the Davy Medal. In 1994, Jeffreys was knighted by Queen Elizabeth II.

Dr. Alec Jeffreys, inventor of the DNA fingerprinting process

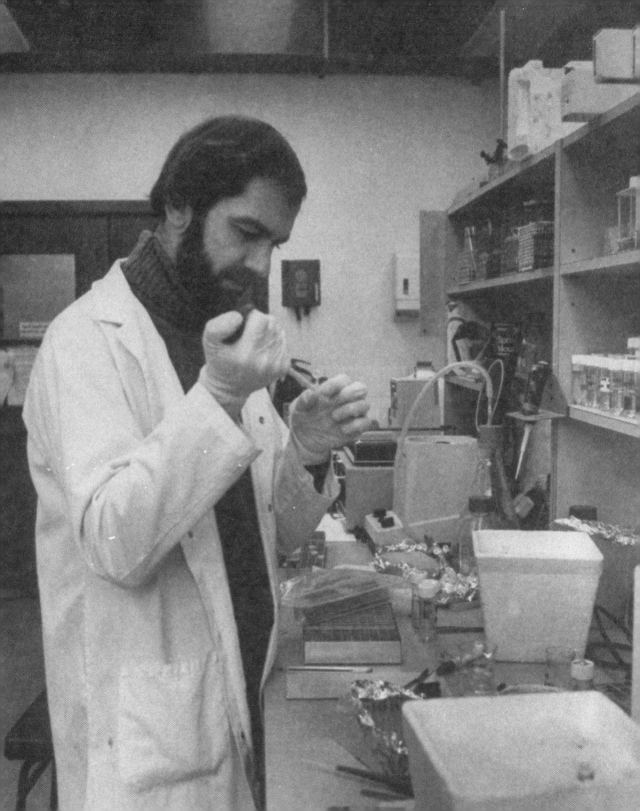

Index